SURGERY OF THE REPRODUCTIVE TRACT IN LARGE ANIMALS

John E. Cox

Department of Veterinary Clinical Science
University of Liverpool

LIVERPOOL UNIVERSITY PRESS

Second edition published 1982 by
LIVERPOOL UNIVERSITY PRESS,
PO BOX 147, LIVERPOOL, L69 3BX

Third revised edition 1987
(reprinted 1990)
Copyright © John E. Cox, 1987

All rights reserved. No part of this book may be reproduced, stored in a retrieval system, or transmitted, in any form or by any means, electronic, mechanical, photocopying, recording, or otherwise without the prior permission of the publishers.

British Library Cataloguing in Publication Data

Cox J.E.
    Surgery of the reproductive tract in large animals. — 3rd rev.ed.
    1. Generative organs — Diseases
    2. Veterinary obstetrics
    I. Title
    636.089'66 SF871
ISBN 0-85323-375-6

Word-processed by the
Composing Unit, University of Liverpool

Printed by
Redwood Press Limited, Melksham, Wiltshire

CONTENTS

                                                  Page

PREFACE

1. **CASTRATION** — 1
   Anatomical considerations - Castration of the horse - Castration of farm animals - Miscellaneous disorders requiring castration

2. **CRYPTORCHIDISM** — 37
   The phenomenon - Approach to a cryptorchid horse

3. **HERNIAS AND RUPTURES IN THE INGUINAL REGION** — 53
   Anatomical considerations - Clinical considerations - Surgical techniques

4. **EQUINE PENIS AND PREPUCE** — 71
   Anatomy - Examination - Clinical conditions - Surgical procedures

5. **BOVINE PENIS AND PREPUCE** — 87
   Anatomy - Investigation - Specific conditions

6. **PREPARATION OF TEASER MALES** — 119
   Indications and pre-operative considerations - Techniques - Post-operative considerations

7. **VAGINAL PROLAPSE** — 127
   Anatomy and pathology - Aetiology and incidence - Treatment

8. **CAESAREAN OPERATION** — 145
   Indications - Alternatives - Techniques - Post-operative course

9. **TRAUMA TO THE FEMALE TRACT** — 171
   Caudal tract - Cranial tract - Vaginal perforations - Cervical Surgery

10. **MISCELLANEOUS CONDITIONS OF THE FEMALE TRACT** — 183
    Ovariectomy - Pneumovagina - Episiotomy - Persistent hymen - Cysts of Gaertner's canals - Cysts of vestibular glands - Vaginal varicosity - Haemorrhage in the mesometrium

APPENDIX I   Restraint and anaesthesia of horses for castration   (i)

APPENDIX II   Restraint and anaesthesia of farm animals for castration   (iv)

APPENDIX III   Pudendal nerve block in the bull   (vi)

APPENDIX IV   Paravertebral anaesthesia in cattle   (x)

INDEX   (xi)

PREFACE TO SECOND EDITION

This book grew out of brief duplicated notes which were given to Final Year Veterinary Students of the Liverpool Veterinary School but it is hoped that the book will provide material of interest and practical value to veterinary surgeons in the field.

Although entitled a surgery book, this book has trespassed from the narrow confines of that art, sometimes for completeness, sometimes for convenience, but mostly because a surgeon's reputation may depend upon his resisting the temptation to wield the scalpel. Emphasis has, therefore, been laid on medical or nursing care or even inactivity when such is more appropriate than surgical intervention.

The sections on anatomy and pathology were provided primarily so that the reader should understand the rationality (or otherwise!) of certain lines of treatment. I have tried throughout to use the Nomina Anatomica Veterinaria (1973) nomenclature, although mostly in English. However, I differed from N.A.V. on occasions and have introduced and defined my own terms where I thought this clinically necessary. In particular, attention is drawn to the term 'spermatic sac', derived by analogy with a hernial sac, and introduced to avoid needless confusion with the term of spermatic cord.

Formal acknowledgements are due to Helen Smyly for the drawings from life in chapter one, to Jennie Smith for the drawings from life in chapter two, (Copyright owned by Unit for Continuing Veterinary Education, Royal Veterinary College, London), to Michael Lynn for the drawings of the inguinal area in chapter three, to Stephen Spencer for most of the drawings in chapter seven and Michael Lomas for the drawings from life in chapters seven and eight. Renee Filbin designed the cover.

It is a pleasure also to acknowledge the advice of many colleagues both from general practice, and from anatomy and surgery departments of veterinary schools across the world, and thanks are due to Paul Neal at Liverpool who read much of the manuscript and made many constructive suggestions and to Ray Ashdown at London for helpful comments on the sections on male anatomy. Errors, as always, remain the responsibility of the author.

Especial thanks are due to Mrs. Anne Harris who cheerfully and expertly turned my crude hand writing into excellent type-written, camera-ready copy and to Mrs. Rosalind Campbell and the staff at the Liverpool University Press for their enthusiastic support and assistance.

Even such a small work as this involved frequent inroads being made into time at home and the patience and forbearance of my wife and young family were beyond the call of duty.

John E. Cox
Leahurst, 1982

PREFACE TO THE THIRD EDITION

The favourable reception accorded the first and second editions of this book has been encouraging. This encouragement, and the continuing demand for the book since the second edition went out of print, have been sufficient to lead to the preparation of this third, revised edition.

Surgery does not seem to make the rapid strides of some other subjects, but new techniques have been described in the literature and new insights have developed from new experiences. Any new edition of a textbook must reflect these advances and help them to find their way into the thinking and armoury of the veterinary surgeon in both referral and general practice. Some of the changes made for this edition have been minor, but the sections on inguinal hernia, equine penile surgery, surgery of the bovine prepuce and vestibular surgery have been rewritten.

Advances in the technology of word-processing have also had an impact and I am grateful to Val Taylor and Sandra Ellams in the University's Composing Unit for their skill in this area. Barrie Edwards contributed new drawings for figs. 4:6 and 10:2. Neil Johnstone of Carson Print Services undertook reprinting of the pictures as most of the originals had been lost in a flood. Neil also redrew figs. 2:1 and 2:2 from the original slides.

I am also grateful for the extraordinary patience shown by Rosalind Campbell and Robin Bloxsidge at the Liverpool University Press and for the support of Michael Clarkson in what has become a veterinary publishing enterprise in the Department of Veterinary Clinical Science.

And I must thank my wife and children who allowed piles of paper to be left in the most inconvenient places and tolerated my invisibility at important times.

John E. Cox
Leahurst
March 1987

This book is dedicated to

my parents

but especially my mother

who wanted to be

a vet

Chapter One
CASTRATION

A. ANATOMICAL CONSIDERATIONS

I. DESCENT OF THE TESTES

The gonad differentiates in the early embryo in the sub-lumbar region but in all the domestic animals the testis migrates from this position through the inguinal canal to the scrotum. This section is concerned with the mechanism and causes of this migration and considers it in some detail as it helps to explain the anatomy of the normal adult and to account for some features of cryptorchidism.

Many of the descriptions of testicular descent in textbooks and elsewhere are inaccurate and based on false premises. Much useful work has been done by Backhouse and Butler (1960) and by Wensing (1968 et passim) on the pig, by Gier and Marion (1969) on the ruminant and by Bergin, Gier, Marion and Coffman (1970) on the horse. All this has provided a sound foundation of knowledge about the domestic species and it is on this that this section is based.

For convenience, the process of testicular descent can be divided into four phases.

(a) <u>Phase One</u> - Formation of the Gubernaculum and Vaginal Process.

Three important facets are involved here,

(1) the structure of the gubernaculum,
(2) the relationship of the gubernaculum to the mesonephric duct and
(3) the formation and structure of the vaginal process.

In the description of events which follows, it will be helpful to refer to Figures 1:1 and 1:2 even though these represent a somewhat later stage of development than phase one.

There develops a cord of mesenchyme which runs from the caudal pole of the gonad through the inguinal region of the abdominal wall to the region ventral and lateral to the primordial pelvis - this cord of mesenchyme is the <u>gubernaculum</u>. The gubernaculum within the abdominal cavity and the gonad soon after differentiation become suspended from the roof of the abdomen in a fold of peritoneum.

The <u>mesonephric duct</u> passes from the mesonephros (which lies <u>lateral</u> to the gonad) through the gubernaculum just caudal to the gonad and so <u>medially</u> towards the developing bladder. This effectively divides the gubernaculum into a shorter proximal part (joining caudal pole of gonad duct to mesonephric duct) and a longer distal part (running from mesonephric duct to blastema of pelvis). The mesonephric duct will eventually differentiate into epididymis and deferent duct.

An evagination of the peritoneum, called the <u>vaginal process</u>, forms in the inguinal region and pushes its way into the extra abdominal part of the distal part of the gubernaculum, separating it into two parts proximally and a third part (that not yet invaded by it) distally. The part of the gubernaculum distal to the invading vaginal process is called the infravaginal gubernaculum. As the vaginal process in cross section is C-shaped (Fig. 1:2B), it divides the remainder of the extra-abdominal portion of the gubernaculum into an outer annular shaped part, called the vaginal part of the gubernaculum, and an inner cylinder shaped part, the gubernaculum proper. The gubernaculum proper is, therefore, suspended within the vaginal process by a fold of peritoneum continuous cranially with the fold of peritoneum which suspends the intra-abdominal part of the gubernaculum and the testis. The C-shaped opening of the vaginal process into the peritoneal cavity is called the vaginal ring, even though it is not a circular opening.

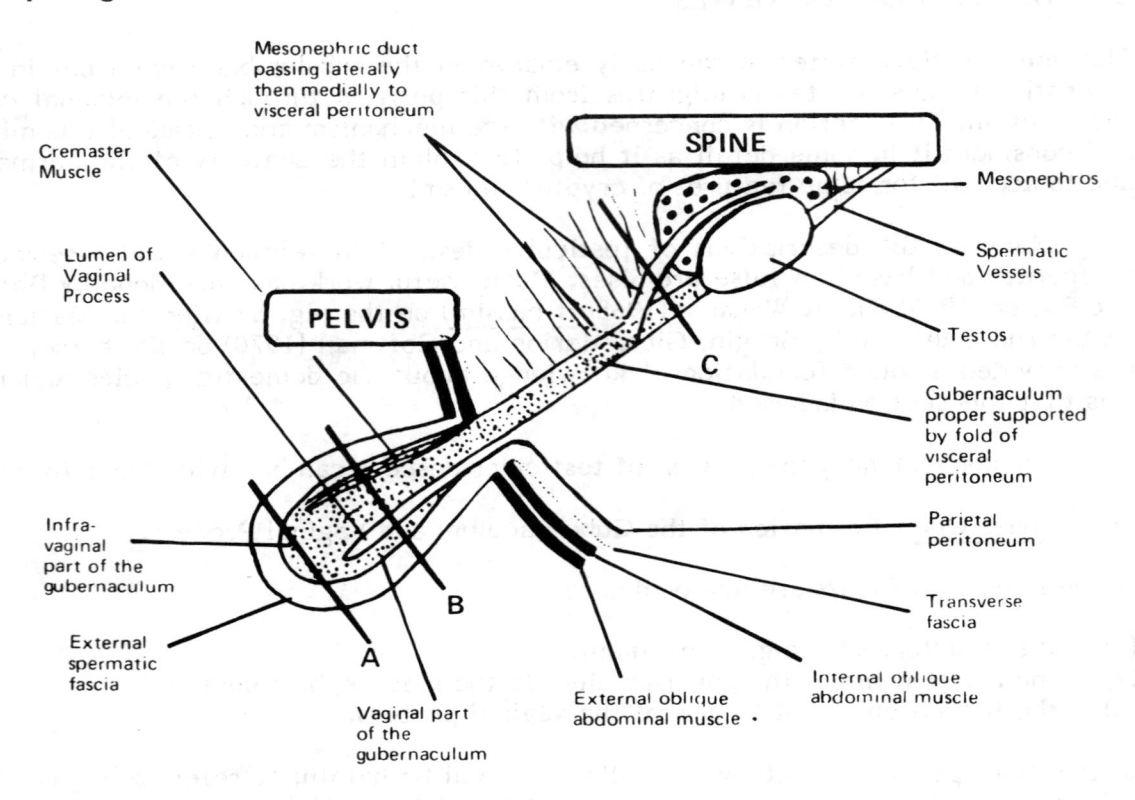

Fig. 1:1 Schematic drawing of the gubernaculum showing its component parts shortly before the mesonephros degenerates. For sections at A, B and C see Fig. 1:2.

(b) <u>Phase Two</u> - Transabdominal Migration

During this phase the testis moves across the abdomen to occupy a position just within the deep inguinal ring. The actual mechanism of this movement is a matter of controversy. At least four factors could be involved here:-

(1) Differential growth between the gubernaculum proper and the lumbar region apparently holds the testis in a fixed position in relation to the deep inguinal ring whilst the metanephric kidney moves cranially with the growth of the lumbar region.

(2) Degeneration of the mesonephros may also play a role in allowing the testis to move caudally. Growth of the metanephros, on the other hand, may push the testis caudally.

(3) Gier and Marion (1969) believe that continued growth of the vaginal process distally maintains tension on an unchanging gubernaculum proper which, therefore, pulls the testis caudally towards the inguinal ring.

(4) Wensing (1968) believes that expansion of the extra abdominal part of the infravaginal gubernaculum plays a role - this expansion draws more gubernaculum through the inguinal canal which then itself expands and draws yet more gubernaculum through the inguinal canal - this movement draws the gubernaculum proper and hence the testis distally towards the inguinal canal.

By the end of this phase the mesonephric duct has become differentiated into head, body and tail of epididymis and into deferent duct. The cremaster muscle develops within the vaginal part of the gubernaculum. Figures 1:1 and 1:2 show the relationship of these structures.

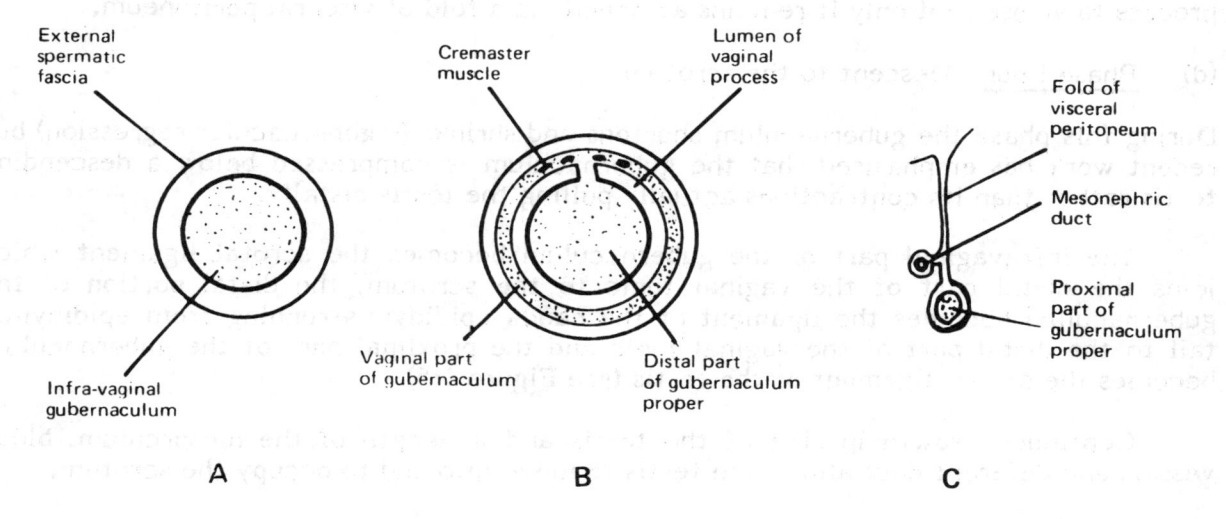

Fig. 1:2 Sections at A, B and C of Fig. 1:1 showing the gubernaculum.

(c) <u>Phase Three</u> - Inguinal Passage

This is undoubtedly the most critical stage of testicular descent. It has often been attributed to shortening of the gubernaculum but this cannot be the cause. Firstly, the gubernaculum does not shorten significantly until after inguinal passage of the gonad. Secondly, the gubernaculum has no strong enough distal connections to make gubernacular contraction meaningful - all that would happen is that the infravaginal part of the gubernaculum would move towards the inguinal canal.

Inguinal passage, therefore, depends on other factors. The gubernaculum proper thickens (= gubernacular outgrowth) and so dilates the inguinal canal to a size through which a testis can migrate. Movement of the testis distally continues as a result of the forces operating in transabdominal migration (see Phase Two above), and in addition, respiratory and other movements of the fetus may increase abdominal pressure and so facilitate expulsion of the testis.

Inguinal passage occurs at about 120 days of pregnancy in the ox, at about 80 days in the sheep and at about 90 days in the pig (Arthur, 1956). In all these species testicular descent is rapid.

In the horse, however, the testis is drawn to the inguinal ring in the normal way, but at about 100-120 days of gestation the fetal gonad begins to enlarge rapidly, precluding descent through the inguinal canal. It reaches its maximum size at about 210-240 days of pregnancy and then shrinks again to weigh only 10 gm at about 300 days of pregnancy. It is now that inguinal passage occurs but this involves distortion of the testis into a long cylinder and takes a considerable time. By the time the foal is born, however, the testes are usually through the inguinal canal.

Much needless confusion arises when the testis is described as passing retroperitoneally across the abdomen and through the inguinal canal. In fact, only in the very early stages of differentiation is the testis retroperitoneal, for it soon hangs from the dorsal body wall in a fold of peritoneum. As, therefore, the testis migrates towards the inguinal canal, it moves across the peritoneal cavity, although encased in visceral peritoneum. As it passes through the inguinal canal it does so through the vaginal process to whose roof only it remains attached via a fold of visceral peritoneum.

(d)   Phase Four - Descent to the scrotum

During this phase the gubernaculum shortens and shrinks (= gubernacular regression) but recent work has emphasised that the gubernaculum is compressed below a descending testis rather than its contractions actually pulling the testis distally.

The infravaginal part of the gubernaculum becomes the scrotal ligament which joins the distal part of the vaginal tunic to the scrotum, the distal portion of the gubernaculum becomes the ligament of the cauda epididymis running from epididymal tail to the distal part of the vaginal tunic and the proximal part of the gubernaculum becomes the proper ligament of the testis (see Figure 1:5).

Continued growth in size of the testis and in length of the mesorchium, blood vessels and deferent duct allows the testis to move into and to occupy the scrotum.

II   THE INDUCTION OF TESTICULAR DESCENT

It has been widely held, though with little good evidence for the domestic animals, that androgens are responsible for testicular descent. However Wensing (1972, 1973a) found no correlation between virilisation and testicular descent in intersex pigs and in a series of experiments on pregnant sows and neonatal dogs, showed that the presence or absence of androgens was without effect on testicular descent (Wensing, 1973b).

More recent and detailed work from this group (see Baumanns and others, 1985) has suggested that androgens are responsible for sexual differentiation and for gubernacular regression. The possibility of a non-androgenic substance of testicular origin being involved in gubernacular outgrowth was suggested by Baumanns and others (1983), but the role of testosterone itself at this phase of testis is equivocal.

Gier and Marion (1969) have emphasised the role of mechanical factors in testis descent, in particular, extension of the vaginal process into the gubernaculum. They believe that closure of the abdominal wall of the early embryo results in an increase in intra-abdominal fluid pressure and that the vaginal process forms by herniation through a weak area in the abdominal wall. Continued later expansion of the vaginal process into the gubernaculum they believe also to be due to increased intra-abdominal fluid pressure. They thus argue that mechanical factors arising outside the gonad/gubernaculum complex are the main factor in promoting descent of the testis. However, in the pig the vaginal process enters the gubernaculum itself (Backhouse and Butler, 1960) not a weak area of the abdominal wall. Gier and Marion's mechanical thesis is, therefore, not entirely satisfactory.

At present there is still no satisfactory answer to the question "what initiates and maintains the gubernacular changes which result in descent of the testis?"

III   ANATOMY OF THE ADULT

(a)   The Scrotum

The scrotum is a sac of skin within which the testes and their coverings lie. In the bull and small ruminants it is long and pendulous having a marked neck, whilst the scrotum of the stallion is more globular with a poorly defined neck. In all these animals the scrotum lies ventral to the cranial end of the pelvis. In contrast, the scrotum of the boar is almost sub-anal, and is less well defined even than the stallions, lying close against the caudal surface of the thighs. The scrotum is supplied with blood from branches of the external pudendal artery. The nerves supplying the scrotum come from the 2nd and 3rd lumbar nerves with a small contribution from the preputial and scrotal branch of the pudendal nerve.

(b)   The Coverings of the Testis

For convenience, the coverings of the testis are held to include the scrotum. Figure 1:3 shows a diagrammatic cross-section of the neck of the scrotum, indicating the layers which are present.

- (i)   The outermost layer is the skin, which is thin and elastic. It is usually without many hairs in the horse but may be covered ventrally with wool in the sheep. It has large sebaceous and sweat glands and is marked centrally in the horse by a pigmented line, the scrotal raphe.

- (ii)   The second layer is the dartos, which consists mostly of fibro-elastic tissue and unstriated muscle. It is only with difficulty that the dartos and skin can be separated except at the neck of the scrotum. The dartos extends dorsally in the mid-line forming the scrotal septum. At the ventro-caudal end of the scrotum, the fibres of the scrotal ligament connect the dartos closely to the vaginal tunic.

- (iii)   The third layer of the scrotum is a series of fascial layers derived from the abdominal muscles. Three layers can be readily identified:-

    - (a)   The external spermatic fascia, which is loosely adherent to the dartos and which is easily separated from (b). In the pig this layer is especially well defined (see Chapter 3)

(b) The underlying cremasteric fascia and the cremaster muscle are part of the fascia of the internal oblique muscle. It is between layers a and b that the surgeon dissects his way when performing a 'closed' castration. The end result of such a dissection is shown in fig. 1:4.

(c) A third layer, derived from the fascia of the transverse muscle and called internal spermatic fascia, is closely adherent to the outer layer of the peritoneum of the vaginal process. The peritoneum itself is the vaginal tunic and is considered in detail below.

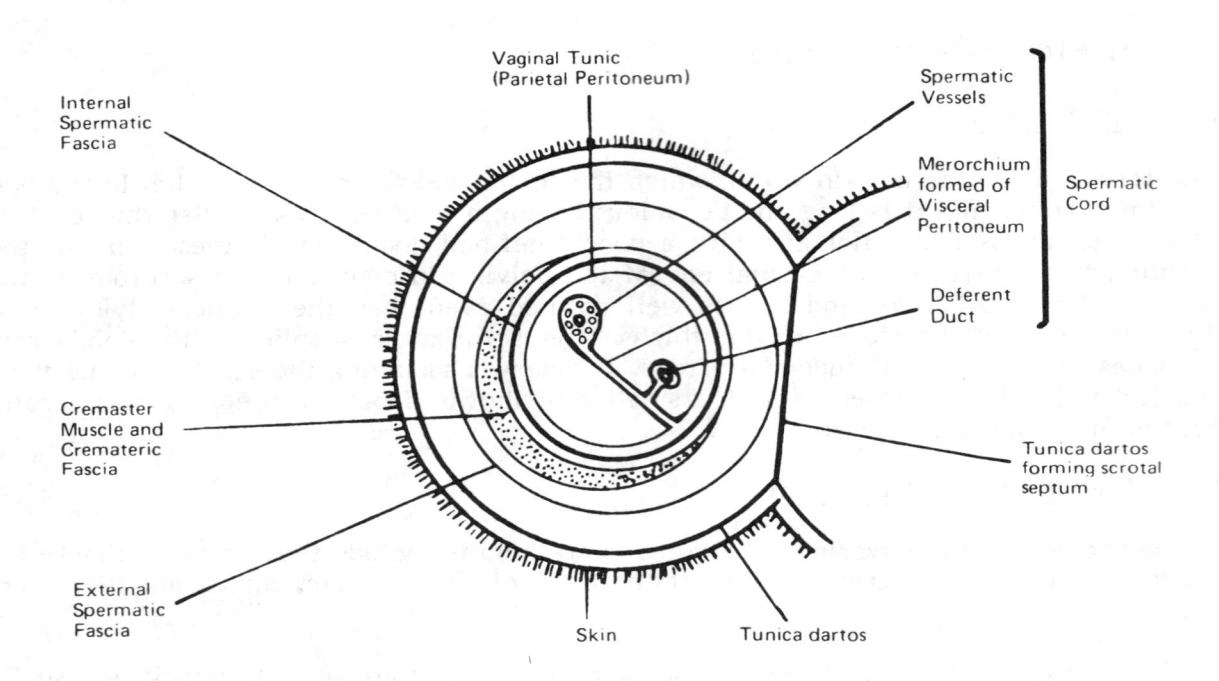

Fig. 1:3 Diagrammatic cross section through neck of scrotum.

(iv) The Vaginal Tunic is a flask-like double layered sac which extends from the deep inguinal ring to the scrotum and consists externally of the parietal peritoneum and internally of the visceral peritoneum lies immediately below the internal spermatic fascia caudally, this peritoneum is reflected and continued as visceral peritoneum over the testis and epididymis distally and as visceral peritoneum of the spermatic cord more proximally (Fig. 1:3). Between the two layers of peritoneum there is a diverticulum of the peritoneal cavity, the vaginal process.

(c)   The Spermatic Cord consists of the structures carried down by the testis in its migration through the inguinal canal and inguinal area. It begins at the vaginal ring where its constituent parts come together, passes through the inguinal canal, and then passes lateral to the penis and ends at the dorsal border of the testis. It takes the form of a wide sheet, the mesorchium, with a thickened cranial border containing the pampiniform plexus of veins and testicular artery, as well as the lymphatic vessels. The caudal edge of the mesorchium is continuous with the vaginal tunic. Medially there is a fold in the mesorchium which contains the deferent duct. Between the two layers of the mesorchium are bundles of unstriated muscle (the internal cremaster of older literature), and also some small blood and lymphatic vessels (Fig. 1:3).

(d)   The Testes are ovoid in form but considerably compressed from side to side. In the ruminants the long axis is vertical (Fig. 5:1) whilst in the horse the long axis is horizontal and in the pig, with its sub-anal scrotum, the long axis runs dorsally and caudally. The greater part of the surface of the testicle is covered by a serous membrane of visceral peritoneum which is reflected at the attached border of the testis to become the parietal peritoneum of the vaginal tunic (see Vaginal Tunic). Beneath this serous coat lies the tunica albuginea of the testis, a strong fibrous capsule. The testis is supplied with blood through the testicular artery which descends in the cranial border of the spermatic cord and is very tortuous near the testicle. It passes along the attached surface of the testis, giving branches to both testis and epididymis, turns around the extremity and runs back along the more convex free border. Its course across the testis is flexuose and it gives off branches which ascend and descend in a tortuous fashion over each surface of the testis. The veins on leaving the testis form a network, the pampiniform plexus, around the coiled artery. These small veins begin to coalesce into larger veins about 10 cm proximal to the testis. These larger veins are well endowed with valves so that blood cannot flow distally. The testicular vein forms by the convergence of these veins high up in the inguinal canal.

(e)   The Epididymis is adherent to the attached border of the testis, overlapping part of the lateral surface (Fig. 1:5). Macroscopically it is divided into a head and a tail, both of which are enlarged compared with the central body. The head is bound to the testis by the visceral peritoneum which envelopes it. The body is less closely bound by the visceral peritoneum, so creating a pocket of peritoneum between testis and body of epididymis lateral called the epididymal sinus. The tail of the epididymis is attached to the caudal extremity of the testis by the proper ligament of the testis and to the vaginal tunic by the ligament of the epididymal tail, both ligaments being contained within short folds of the visceral peritoneum. The tail is continued as the deferent duct which ascends through the inguinal canal enclosed in a medial fold of peritoneum detached from the mesorchium. Alongside the deferent duct lie the deferent artery and vein.

In the bull and ram, the head of the epididymis is largely dorsal but it extends over the cranial surface of the testis, the body runs down the medial surface of the testis (not the lateral as stated in Sisson) and the tail lies ventral to the testis. In the stallion the epididymis lies dorsal to the testis with its head cranially and its tail caudally whilst in the boar the epididymis lies on the cranio-dorsal surface of the testis with the tail caudally and dorsally. In the ruminants and horse, the tail is readily palpable and in the adult pig the tail is often clearly visible. The head can be appreciated as an extension or cap to the testis. The body of the epididymis is difficult to palpate satisfactorily in any species.

(f) The Spermatic Sac This is a term used throughout this book to describe the whole of the cremaster muscle and fascia, internal spermatic fascia and the vaginal tunic (Fig. 1:4). It is the distal part of this structure which is dissected free when a 'closed' castration is performed. It is unfortunate that the term spermatic cord is often synonymously with it, even in Nomina Antatomica Veterinaria. In this book, 'spermatic cord' is used sensu strictu, and 'spermatic sac' in the sense given here. By analogy with a hernial sac (see p.58), the spermatic sac has a ring (the vaginal ring), a neck (that part contained within the inguinal canal), a body (that part between superficial inguinal ring and the most dorsal part of the epididymis) and a fundus (that part containing the testis and epididymis).

IV GENERAL COMMENTS

The attachments of the testis and adnexa and the surgical significance of the blood supply can now be appreciated.

The attachments which hold the organs in position are as follows. Firstly, joining the distal part of the vaginal tunic to the scrotal fascia there is the scrotal ligament. This is most noticeable when doing a 'closed' castration (p.13) or repairing an inguinal hernia (p.59) but as it contains no significant blood vessels and is not very strong it may be safely and easily snapped. Secondly (Fig. 1:5), there are the ligaments of the epididymis which, by running from the tail of the epididymis to the testis (= proper ligament of the testis) and from the tail of the epididymis to the distal vaginal tunic (= ligament of the tail of the epididymis), anchor the testis to the distal vaginal tunic. Although this epididymal ligament complex is derived from the gubernaculum, with age it tends to become stronger and so less readily snapped. Thirdly, there is the mesorchium and its divisions suspending the epididymis and deferent duct.

The most likely source of haemorrhage post-castration is the spermatic artery - bleeding from the distal testicular veins is unlikely because they are well endowed with valves. In the older animal, bleeding from the deferent duct vessels is possible as also is bleeding from the scrotum. The large vessels in the scrotum run in the mid-line septum, incision of which is, therefore, undesirable. Another possible source of haemorrhage are vessels which may run in the fold of peritoneum forming the mesorchium independantly of the pampiniform plexus.

It is especially important to realise that the lumen of the vaginal process is a diverticulum of the peritoneum cavity. The surgical significance of this anatomical fact cannot be over-emphasised. It is thus possible for the contents of the abdominal cavity to pass through a ready-made hole in the body wall, viz. the vaginal ring. Moreover, if an incision is made through the scrotal coverings into the lumen of the vaginal process, this opens up a route for material such as intestines to pass from the abdominal cavity into the outside world or for material such as infective agents to pass from the outside world into the peritoneal cavity. The consequences of this anatomical fact can be disastrous. It should be noted that in the adult human the vaginal ring and the neck of the vaginal process are usually obliterated and the above remarks do not apply.

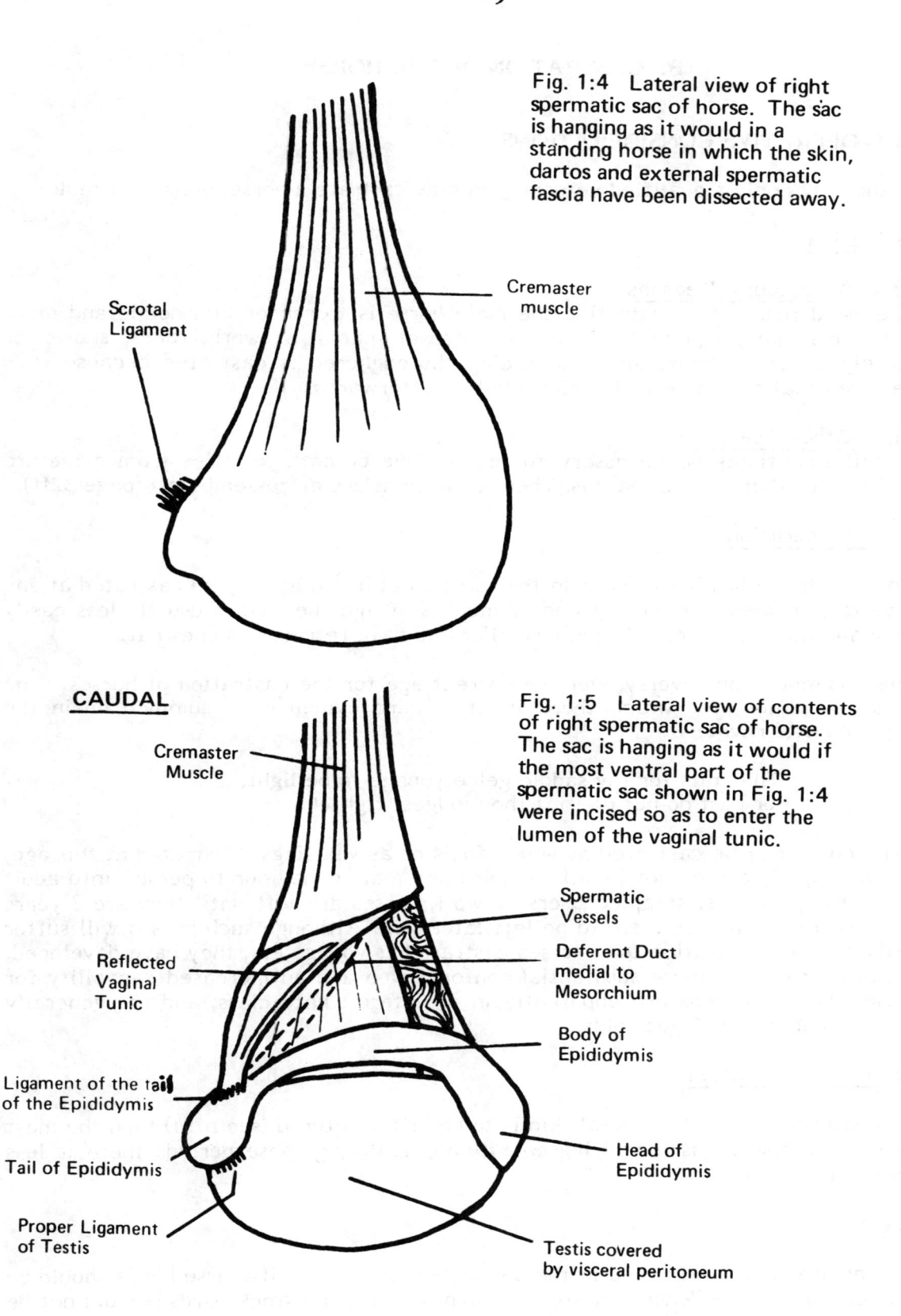

Fig. 1:4 Lateral view of right spermatic sac of horse. The sac is hanging as it would in a standing horse in which the skin, dartos and external spermatic fascia have been dissected away.

Cremaster muscle

Scrotal Ligament

CAUDAL

Fig. 1:5 Lateral view of contents of right spermatic sac of horse. The sac is hanging as it would if the most ventral part of the spermatic sac shown in Fig. 1:4 were incised so as to enter the lumen of the vaginal tunic.

Cremaster Muscle

Reflected Vaginal Tunic

Spermatic Vessels

Deferent Duct medial to Mesorchium

Body of Epididymis

Ligament of the tail of the Epididymis

Tail of Epididymis

Head of Epididymis

Proper Ligament of Testis

Testis covered by visceral peritoneum

B. CASTRATION OF THE HORSE

I   PRE-OPERATIVE CONSIDERATIONS

(a)   In the U.K., only a qualified veterinarian may castrate a horse, donkey or mule.

(b)   Indications

   (i)   <u>Behavioural Reasons</u>
   The usual reason for castrating the male horse is to render him docile and more easily managed, especially in the presence of mares, at work, races, shows, or merely at grass. Sometimes the racing thoroughbred is castrated because it is believed that the horse will perform better afterwards.

   (ii)   <u>Other Reasons</u>
   It will sometimes be necessary to remove one or both testicles from a mature stallion for therapeutic reasons. These conditions are discussed below (page 32ff).

(c)   Age at Operation

As the normal horse has both testes in the scrotum at birth, he may be castrated at any age. However, between 4-6 months and 12 months of age the foal is usually less easily restrained and the testes are relatively smaller and have retracted somewhat.

There is much controversy over the correct age for the castration of horses. The concensus of opinion, for which there is little hard evidence, is summed up in the following medieval rhyme:-

   "Thy colts for thy saddle gelde younge to be light;
   For cart do not do so if thou judgest righte".

Thus, pony colts may be castrated as young foals or as yearlings. Castrated at this age, they recover rapidly and do not usually acquire any male behaviour to persist into adult life. In contrast, hunters, steeplechasers or work horses are left until they are 2 years old and many owners prefer them to be left later still. Although such horses will suffer more setback from castration than horses castrated as youngsters, they have developed, and apparently retain, a more substantial conformation and an increased capability for hard work. Thoroughbreds are apparently in an intermediate class, and are generally castrated as yearlings or 2 year olds.

(d)   Season for Operation

If, as is usually the case, the scrotal skin is to be left unsutured (see p.14) then the most suitable time to operate is the spring or autumn, as during these periods there is less likelihood of fly strike.

(e)   Place

If operations are to be performed in the open, then a clean, well-grassed area should be chosen. Areas where the density of stock is high, e.g. around stock yards, should not be chosen due to the increased danger of wound contamination from dust and of tetanus infection.

(f) <u>Pre-operative Examination</u>

Castrations should not be performed in stables where diseases such as equine influenza or strangles are present, nor in horses in poor bodily condition.

A careful examination of the scrotal area should be made to ensure that both testes are present and that there is no inguinal hernia. The scrotum can be palpated by the surgeon standing close to the horse's chest with his head close to the horse's withers and palpating with his right hand (Fig. 1:6). The scrotum can be viewed from behind by lifting the tail. In the normal horse the left scrotum is usually slightly more pendulous than the right, but any marked discrepancy in size suggests the possibility that an inguinal hernia is present. The owner should, in any case, be asked if an inguinal hernia has ever been present and if a definite history of the previous presence of this condition is found, then it is imperative that a 'closed' castration be performed. Some authorities advocate examination of the size of the vaginal ring <u>per rectum</u> to estimate its size and, therefore, the likelihood of intestinal prolapse, but <u>factors</u> other than the size of the ring contribute to intestinal prolapse (see p.63) and such a procedure may only pick out a small proportion of externally normal horses which will later develop eventration.

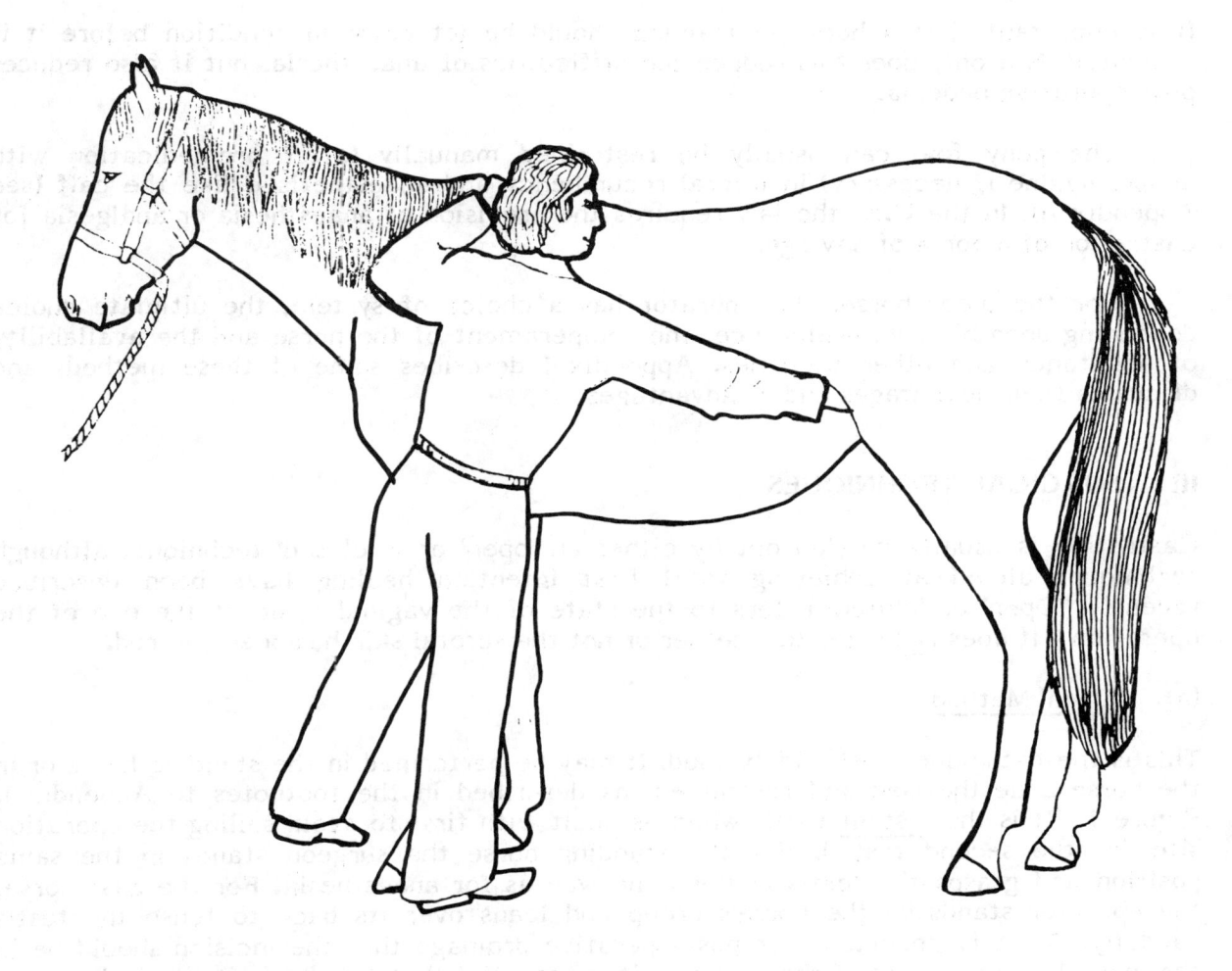

Fig. 1:6 Showing palpation of inguinal region being carried out by a surgeon who is standing safely by the fore legs and is keeping his head well up.

(g) Prophylactic Measures

Tetanus anti-serum should be administered prophylactically to every horse which is to be castrated - this is only omitted if a definite history is available of the previous administration of toxoid and, even then, a further dose of toxoid immediately prior to surgery is indicated.

The <u>local</u> administration into the castration wound of sulphonamide powders, antibiotic creams or antibiotic solutions has nothing to commend it - these materials can behave as foreign material and increase the tissue reaction, so delaying healing.

The administration of antibiotics <u>systemically</u> should always be made <u>prior</u> to surgical intervention so that the exposed tissues are saturated with antibiotic at the time of maximum exposure risk, thus supplementing the animal's natural defence mechanisms. The intravenous administration of crystalline penicillin (not less than 5 mg/Kg) immediately after induction of anaesthesia is the method of choice because the two major pathogens of the horse (Streptococci and Clostridia) are penicillin-sensitive.

II RESTRAINT AND ANAESTHESIA

It is important that a horse in training should be let down in condition before it is castrated. Not only does this reduce the difficulties of anaesthesia, but it also reduces post-operative oedema.

The pony foal can usually be restrained manually (after premedication with acepromazine if necessary) in lateral recumbency and anaesthetised like the calf (see Appendix II). In the U.K. the law requires the provision of anaesthesia or analgesia for castration of a horse of any age.

For the older horse, the operator has a choice of system, the ultimate choice depending upon his own preference, the temperament of the horse and the availability of assistance and other facilities. Appendix I describes some of these methods and discusses their advantages and disadvantages.

III SURGICAL TECHNIQUES

Castration is usually carried out by either an 'open' or a 'closed' technique, although techniques aimed at achieving total first intention healing have been described recently. 'Open' or 'closed' refers to the state of the vaginal tunic at the end of the operation - it does <u>not</u> refer to whether or not the scrotal skin has been sutured.

(a) 'Open' Method

This is the technique most widely used. It may be performed in the standing horse or in the horse anaesthetised and restrained, as described in the footnotes to Appendix I, Figure 1. (It is the <u>distant</u> testis which is dealt with first to avoid soiling the operation site for the second testis). For the standing horse the surgeon stands in the same position and grasps the testis in the same way as for anaesthesia. For the cast horse, the operator stands by the horse's croup and leans over its back to tense the testis caudally. As it is important for post-operative drainage that the incision should be in the most dependant part of the scrotum, it is essential that the testis is grasped so that the median raphe of the scrotum and the long axis of the testis lie parallel to one another and that the incision is made along the most prominent part of the scrotum.

The incision should be made from the front of the scrotum to the caudal end (it's safer that way) and should be of a generous length (again to allow adequate post-operative drainage) and should run about 3-5 cm from and parallel to the median raphe. The incision is made to pass through skin, dartos, spermatic fascias and vaginal tunic so that the lumen of the vaginal process is entered and the testis readily protrudes. The testis should now be grasped and pulled distally to expose the spermatic cord well proximal to the epididymis. The view obtained is that illustrated in Fig. 1:5. A good emasculator (see (e) below) is placed on the cord as proximally as possible with the crushing edge proximal and the cutting edge distally and closed securely until the testis falls off. Authorities differ as to how long the emasculator should remain in place, recommendations varying from 30 sec. to 4 min.

The whole procedure is repeated on the other testis.

The advantages of the technique are its speed and simplicity and the fact that it can be carried out without touching anything which remains inside the horse. The disadvantages are that the risk of herniation of intestine is present, that infection readily gains access to the inguinal area and can spread into the abdominal cavity, and that leaving most of the sac behind allows the possible development of champignon, scirrhous cord and cystic ends (see pp.19-20 below). Nevertheless, the method is the one which is most widely used. As Vaughan (1978) points out, it is still necessary to insist upon safe, clean surroundings and not to surrender the patients welfare on the sacrificial altar of expediency.

(b) The 'Closed' Method

Although it is possible to perform a closed operation on a horse restrained in left lateral recumbency, the operation is most efficiently carried out with the horse in dorsal recumbency, a position easily maintained in the field by placing a bale at each shoulder. In order to perform a 'closed' operation, one of the longer acting anaesthetic agents such as chloral hydrate, thiopentone/halothane (or chloroform) or 'Immobilon/Revivon' are necessary. As a ligature is to be buried in the animal the standard of surgical cleanliness of operation site and operator must be high.

The testis is tensed into the scrotum as for the 'open' method and a careful incision is made through the skin, dartos and external spermatic fascia only - the vaginal tunic, cremasteric fascia and internal spermatic fascia are not incised. As with the 'open' method, it is necessary that the incision be in the most dependant part of the scrotum and, therefore, that it be aligned with the median raphe. By blunt dissection the spermatic sac is dissected free. The external spermatic fascia strips away cleanly from the spermatic sac if, and only if, the plane between external spermatic and cremasteric fascial planes is found. The only obstruction to the correct dissection is the scrotal ligament (the remains of the infravaginal part of the gubernaculum which binds vaginal tunic to external spermatic fascia caudally). The view obtained is that shown in Fig. 1:4. A ligature of 5 metric synthetic absorbable suture material is anchored through the external cremaster muscle and tied securely round the neck of the sac reasonably high up (Fig. 1:7).

Fig. 1:7 Diagrammatic cross-section through neck of spermatic sac showing ligature

The sac is transected not less than 2 cm distal to the ligature with an emasculator. (The use of an emasculator alone without ligation will not provide secure haemostasis - see Cox, 1984).

The 'closed' method has the advantages that it prevents the possibility of intestinal prolapse after castration and that, as a length of spermatic sac is removed, there is less risk of post-operative complications such as cystic ends (see p.20 below). Moreover, the peritoneal cavity is sealed against infection. The disadvantages of the method are that it requires more surgical and anaesthetic time (and cannot, therefore, be satisfactorily performed under quick-shot anaesthetic techniques) and that it involves placing a ligature deep in the inguinal region (and, therefore, a high degree of surgical cleanliness is required). In the author's opinion, however, the advantage of security from herniation outweigh the disadvantages of the longer time and greater care required.

(c) Techniques achieving First Intention Healing

The introduction of readily available suture materials at the turn of the century and the possibility of achieving first intention healing after castration led some operators to attempt closure of all tissues incised, but the poor quality of much of the available suture material gave problems and the attempts were abandoned. It then became an axiom of castration that on no account was the scrotal incision to be sutured.

In 1951, however, Formston described a technique for foals which involved performing a closed castration and suturing the scrotal skin afterwards and gave good results. In 1964, Roberts reported that, provided due attention was paid to haemostasis and to surgical cleanliness, no untoward sequelae followed closure of most of the scrotal wound in adult stallions with matress sutures. Since that date, there have been a spate of reports advocating techniques which acheived first intention healing (Lowe and Dougherty, 1972; Palmer, 1984; Rutgers and Merkens, 1983; Barber, 1985).

Since 1973, the author has performed closed castrations as described above in Section (b) and sutured the scrotal skin with a continuous mattress suture, using synthetic absorbable suture materials. The dead space left by removal of the distal part of the spermatic sac, including the testis, has not been closed by suturing but has merely had the air squeezed out of it (Cox, 1984). The procedure takes 8-12 minutes from first incision to completing closure on the second side. The cosmetic results have been highly acceptable - there is minimal oedema and rapid convalescence. Experience suggests that the infliction of minimal trauma is an important contributory factor to minimal oedema and, therefore, recognition of the plane of separation of spermatic sac from external spermatic fascia is important. Other factors include the use of unreactive suture material and a high degree of surgical cleanliness. The technique is applicable to field conditions where it can give results as good as those achieved in a hospital environment.

The <u>advantages</u> of this latter technique are those of a 'closed' castration plus the more rapid healing and convalescence which follows because secondary infection is totally excluded. The <u>disadvantages</u> of the technique are again those of a 'closed' castration, in particular the need to be scrupulously clean in the course of performing the operation. It is imperative that any one undertaking a 'closed' operation or intending to suture the scrotal skin should achieve a far higher degree of surgical cleanliness than is necessary for 'open' castrations.

(d) <u>Other techniques - not recommended</u>

(i) <u>The use of a ligature for haemostasis in the 'open' method</u>

A common variant of the technique described under (a) above is to separate the exposed cord into a cranial 'vascular' portion, containing the spermatic vessels and a more caudal 'non-vascular' portion containing the duct and fold of vaginal tunic. This separation is easily accomplished by pushing a finger through the mesorchium. The non-vascular part is severed with an emasculator or, in the young horse only, simply cut through. A ligature of absorbable suture material is then tied round the blood vessels as far proximally as possible and the vessels severed distally. (A variant on the use of the ligature here is to use torsion forceps which are designed to grip the cord and are then twisted round and round until the cord snaps - torsion forceps are rarely used now).

The <u>disadvantage</u> of this approach to haemostasis over the use of the emasculator along is that it involves handling the spermatic cord and so increases the risk of septic organisms being introduced. Nevertheless, in the <u>adult donkey</u> ligation is recommended because of the increased risk of haemorrhage in this species.

(ii) <u>The use of clams</u>

Before the advent of the ecraseur and emasculator, a common way of achieving haemostasis was by the use of wooden or metal clams applied to the exposed cord. According to one's preference, it was possible to remove the testis and epididymis distal to the clam by the scalpel or a hot knife. Moreover, authorities differed as to whether or not a caustic ointment should be applied to the cord along with the clam or after their removal. In general the clams were left in position for 24 hours before being removed.

The method is still occasionally used but has no advantage over a good emasculator and has the <u>disadvantage</u>, especially if caustics are used, of producing a marked fibrous reaction in the cord and of delaying healing.

(e) The Emasculator

The emasculator is an instrument designed to carry out two tasks, viz. to crush the spermatic cord or sac proximally and to cut it distally. Its use, therefore, means that haemostasis and cutting are performed by one instrument. In particular, it is intended to obviate the need to use a ligature.

Since its introduction the emasculator has been modified in a number of ways and a large number of different types are available today. If one is to rely upon it and upon it alone for haemostasis, then it is imperative that it should do its job effectively. Accordingly, not only should the correct model be chosen, but it should be looked after most carefully.

Of the models available, and known to the author, the 'Serra' emasculator in the form modified by Bertschy, or with its handles modified by Crowhurst, is undoubtedly the best. As it is closed round the cord it, in turn, grips, crushes and then cuts. In particular, the curve of the crushing blades is such as to crush the whole cord almost simultaneously and the grooves on the blade run proximally/distally, obviating the possibility of the crushing blade cutting the cord. In the Verboczy emasculator, the ridges on the crushing blade on the proximal side run across the instrument and can cut the cord instead of crushing it. Moreover, the straight nature of the cutting surfaces means the cord is crushed unevenly and, in use, the instrument will sometimes be found to cut the cord distally before the cord is satisfactorily crushed. In others, e.g. the Hausman, there are no two surfaces which crush together - these types are unsatisfactory. The Reimer emasculator has a separate cutting assembly - the cord or sac is crushed first and then a separate handle and blade are brought into use to sever the testis distally. However, the jaws are relatively smaller than the Serra and it is not as easy to get the large spermatic sac of an adult stallion within them.

Proper maintenance of the emasculator is also important. After each use, the instrument should be dismantled and each part thoroughly cleaned and smeared with petroleum jelly prior to the instrument being reassembled. Ideally, therefore, the model chosen should be capable of being readily dismantled and re-assembled. Of those mentioned above the Serra is the easiest to dismantle.

IV COMPLICATIONS

(a) Haemorrhage

Most cases of post-castration haemorrhage are primary and, therefore, result from failure to achieve adequate haemostasis. The possible sources of haemorrhage are the scrotal skin, the deferent vessels and the testicular artery. The testicular veins are well endowed with valves so distal venous bleeding is almost impossible. In a 'closed' castration, bleeding from vessels in the spermatic fascia or the cremaster muscle is also possible.

Slight dripping from the wounds a few minutes after the operation gives no cause for worry. If the drips are almost running into one another or if blood is streaming from the wound and shows no sign of slowing down then action should be taken. In such a case it is tempting to pack the wound with a sterile gauze (not cotton wool) for 24 hours but there is the risk of introducing gross infection and of the animal later developing other complications. It is preferable to identify the bleeding point or points and deal with the haemorrhage by direct haemostasis, even if this means re-anaesthetising the animal.

The cut edges of the vaginal tunic should first be identified and held by tissue forceps. A pair of haemostats is then introduced into the lumen of the tunic and the spermatic vessels grasped and dealt with as necessary.

A rare complication of haemorrhage following castration is blindness. The blindness is usually sudden in onset, total and irreversible and due to ischaemic optic atrophy and the development of a chorio-retinopathy.

Obviously prevention of haemorrhage is better than cure. Adequate restraint and anaesthesia help to achieve good haemostasis by preventing struggling or movement. The haemostatic technique used should also be adequate for the age of the animal.

If haemorrhage occurs following one of the techniques using first intention healing, then a haematoma develops in the scrotum above the sutures in the scrotal skin. The horse should then be given three further daily injections of penicillin and on the third day the scrotal skin sutures removed (under a general anaesthetic if necessary) and the haematoma evacuated by a gloved hand. The scrotal wound is then left to heal by second intention.

(b)   Eventration

Following an 'open' castration either omentum or bowel may prolapse through the vaginal ring and scrotal incision. Both eventualities are discussed further in Chapter Three.

(c)   Non-specific swelling and infection

   (i)   Swelling

Some scrotal swelling almost invariably develops after castration in the horse, probably due to oedema which characterises any wound, surgical or otherwise, in this species. In the uncomplicated case, the swelling reaches a maximum at 4 days and then subsides. When horses in training are castrated, the post-operative swelling is usually greater than that experienced in horses on grass.

Occasionally inguinal swelling develops when fluid accumulates behind a scrotal incision which has sealed over too quickly following a castration in which the scrotal wound was unsutured. The fluid is probably inflammatory in origin. The swelling is easily relieved by re-opening the scrotal wound with gloved fingers (the hand is gloved, not to prevent the introduction of infection into the inguinal area, but to avoid soiling the veterinarian's hand with bacteria which may be pathogenic to other patients).

   (ii)   Infection

Some infection almost invariably finds its way into the scrotal wound which is left unsutured and this infection probably contributes to the oedema. Flicking of the tail undoubtedly encourages infection as does residence in a strawed box or a dusty yard. In many cases, the infection does not become established probably because drainage through the scrotal wound occurs continuously. In other cases, infection may become established either because a large number of bacteria are introduced into the wound due to faulty technique or dirty surroundings or because the scrotal wound becomes sealed over too quickly or both.

Traditionally, this infection and the oedema which develops are controlled in several ways. <u>Firstly</u>, a long scrotal incision is made with the expectation that this has less chance of healing over quickly. <u>Secondly</u>, the scrotal incision is made in the most dependant part of the scrotum, thus ensuring that there are no pockets of skin in which infection can become established. <u>Thirdly</u>, the horse is exercised, either by half-an-hour enforced walking or by being turned loose into a paddock, for 10 days after castration. (Exercise helps, not only by controlling oedema, but also by encouraging movement of the edges of the scrotal wound and so delaying slightly complete closure.)

Occasionally, however, and perhaps in spite of the measures just outlined having been taken, infection does become established after castration. In such a case oedema is usually marked and may be serious enough to cause preputial and penile swelling (p.90). The animal is usually depressed and inappetent and there is marked staining of the inner aspects of the thighs with discharge from the inguinal region.

Treatment must be directed at releasing the infection ('where there's pus, let it out') so a gloved hand should be used for several days to keep the scrotal wound open and the infection draining away. A course of antibiotic therapy should be instituted. Should the oedema be so severe as to have caused prolapse of the penis, then attention should be directed at it as outlined on page 93.

Although recovery may appear to be complete, there is some evidence (Goulden, personal communication) to suggest that rheumatic heart disease may follow streptococcal infections in the horse in much the same way as rheumatic heart disease in humans follows upper respiratory tract infections with streptococci. The need for a veterinary surgeon to avoid post-operative sepsis by his technique is thus underlined.

(iii) <u>Abscess</u>

Simple Abscesses usually develop within one or two months and either burst spontaneously soon after developing or can be lanced. They may be accompanied by severe malaise from toxaemia. Provided drainage is maintained, healing is usually rapid and complete.

(d) <u>Specific Infections</u>

(i) <u>Champignon</u>

This is said to be due to infection with a Lancefield Group C Streptococcus and is characterised about 14-21 days after castration by the appearance of a swelling in the inguinal region, a purulent discharge and general malaise. It is associated with the formation of exhuberant pink granulation tissue (Degive, 1875), which may ultimately protrude from the scrotal wound, mushroom-shaped (hence the name).

The only successful treatment is removal of the affected tissue through an incision lateral to the original. This is not usually difficult as the unhealthy tissue is about the end of the remains of the spermatic sac and dissects out cleanly. The wound is, of course, left open to heal by second intention. (The author has never seen a classical "Champignon" and wonders if the widespread use of clams before the invention of the emasculator was a contributory factor in the development of champignon).

(ii) Suture Abscess

The author has now seen 6 cases of suture abscesses, all but one in horses in which the ligature material was multifilament nylon with a plastic sheath. The animals presented years after castration with a discharge from the scrotum and very little swelling. Inspection under general anaesthesia showed a discharging sinus with a small tract of varying depth at the bottom of which lay the ligature. Some thickening of the spermatic sac was apparent and palpation occasionally promoted the discharge of pus. Excision of the distal part of the spermatic sac and the skin surrounding the sinus was possible. It is important to excise all the thickened part of the sac and get into the healthy atrophied spermatic sac proximally. Sometimes, however, the pathological process has extended up into the groin and surgical removal is not effective. A large skin deficit is created by the surgery and considerable post-operative swelling characterises the prolonged recovery.

(iii) Scirrhous Cord

The classic scirrhous cord is a botriomycotic lesion consisting of extensive fibrous tissue and micro-abscesses, and caused by a low grade infection with staphylococcal organisms. The lesion does not usually become apparent until several years after castration and is characterised clinically by a painless, sometimes large, swelling in the inguinal area with many openings in the skin discharging pus. Occasionally, the disease process will be found to have spread through the inguinal canal into the abdomen and the horse may be showing loss of condition - in such a case treatment is hopeless.

The only possible treatment is surgical excision but the prognosis is always guarded and convalescence prolonged and surgery should not be undertaken lightly. A pre-operative course of appropriate antibiotic for 10 days may help to reduce the size of the lesion. Removal is extremely difficult as the healthy tissue and affected tissue have no clear-cut boundary and the removal of large amounts of tissue is often necessary.

The wound will require irrigation perhaps for several weeks before granulation tissue fills the cavity and the wound heals.

As with champignon, the author has never seen a case of classic scirrhous cord and wonders again if the widespread use of clams and of caustic was a significant factor in the development of scirrhous cord.

(e) <u>Fly strike</u>

The wisdom of avoiding the fly season when doing open castrations was brought home to the author recently. A Shire horse had been castrated in the summer, stood up after anaesthesia, prolapsed his intestines, was re-anaesthetised to have the guts poked back under the scrotal skin and was allowed to recover before being sent to the Large Animal Hospital. At the Hospital 15 feet of small intestine was resected and healthy bowel returned to the abdomen and the lumen of vaginal tunic closed over. The scrotal wound was left to heal by second intention and a week later the site was crawling with maggots - the horse recovered from that, too!

(f) <u>Cystic Ends</u>

This complication may develop or become apparent only months or years after castration. It is characterised by the development of cysts at the end of the spermatic sac which distend the scrotum, giving the animal the appearance of having a scrotal testis. On palpation, however, the swelling will be found to be due to fluid. In some cases, the fluid can be squeezed up into the abdomen and in other cases it cannot - this will depend upon whether or not the cyst or cysts are continuous with the lumen of the vaginal process. The owner may report that the swelling has developed suddenly and there may be concern that it is due to an inguinal hernia. However, inguinal hernia in the established gelding is extremely rare and the soft fluctuating swelling of the cysts is characteristic.

The presence of the cysts does not inconvenience the horse and, unless the owner wishes the blemish to be removed, there is no need to do anything about it. With the horse under general anaesthesia and in dorsal recumbency, the remains of the spermatic sac and the cysts can be dissected out and removed. There are usually considerable adhesions between the cysts and sac on the one hand and the scrotal skin on the other so the dissection is not particularly easy and post-operative swelling may be worse than the original blemish.

(g) <u>Removing Penis instead of Testis</u>

The author has seen a horse which was operated upon in a dark loose box and in which the penis was grasped and incised instead of the testis. Pearson (personal communication) reports several cases in which castration has been attempted in the foal and in which the penis has been mistaken for a testis and been totally ligated and removed with occlusion of urethra followed by rupture of the bladder. One such foal survived for almost a month because urine escaped from the peritoneal cavity through the lumen of the vaginal tunic on one side.

These cases illustrate that even an apparently simple operation requires attention to detail.

C. CASTRATION OF FARM ANIMALS

I PRE-OPERATIVE CONSIDERATIONS

(a) <u>Advantages and Disadvantages</u>

Castration of farm animals has probably been practised since antiquity, but the need to do so has recently been questioned on scientific, economic and humanitarian grounds. In particular, there has been a change to faster maturing breeds of animals and, as these may reach slaughter weight before puberty is complete, it is much more difficult to justify castration on the grounds that it produces quieter, more easily managed stock. The introduction of the horse to agricultural work, about 1775, and later the advent of the tractor has, in developed countries at least, obviated the demand for castrated cattle for draught work. An excellent summary of the advantages and disadvantages of castrating farm animals was made by Robertson (1966) and an interesting article on behavioural aspects of the subject was written by Kiley (1976).

The <u>advantages</u> claimed for <u>castration</u> are as follows:-

1. Indiscriminate mating is prevented, so allowing controlled systems of breeding to be employed by the farmer or the state.

2. In the castrated animal there is increased deposition of fat, both between and within muscles, and this has often been held to be advantageous.

3. Carcass quality is improved in the castrated animal because of the absence of taint, especially in the pig, but also in cattle.

4. Aggressiveness and libido are decreased by castration and so management is facilitated. Males and females may be safely housed or run together.

The <u>advantages</u> of <u>leaving animals entire</u> are as follows:-

1. Selection of males for breeding would no longer be hindered by loss of potentially good stud animals and, therefore, improved selection would result.

2. In the entire animal there is less fat and more protein per animal and, at present, the consumer prefers leaner meat.

3. Growth weight gains and efficiency of food conversion are improved and there is no check in weight gain such as occurs in the period immediately following castration. Total protein output for food input is, therefore, increased.

At present the situation is nicely balanced and although, in a given management and feeding and marketing regime, castration is advantageous, in another it can be dispensed with. The article by Robertson provides interesting reading material in this respect.

(b) <u>Legislation</u>

However, several pieces of legislation in the U.K. may affect a farmer's attitude to castration. It is not possible to be detailed about this legislation as it changes rapidly. Briefly, it falls under three headings:

(i) <u>Licensing Regulations</u> These regulations prevented the keeping of entire males over certain ages unless the animal is intended for and is licensed as suitable for stud purposes. Most of these regulations have been suspended, except in the case of bulls for A.I.

(ii) <u>Fatstock Guarantee Schemes</u> These schemes give guaranteed prices for certain classes of animals. Normally entire males are excluded but at the time of writing the scheme only applies to sheep - entire ram lambs are not ineligible for grading unless they show ram like characteristics or are slaughtered after approximately January 1st following the year of their birth.

(iii) <u>Cruelty to/Protection of Animals/Anaesthesia</u>
Bulls, male goat, and male pig, may only be castrated without an anaesthetic and by a lay-person up to 2 months of age. The corresponding age for male sheep is 3 months. Beyond this age, anaesthesia or analgesisa must be provided and the operation must be carried out by a qualified veterinary surgeon or practitioner.

(iv) <u>Miscellaneous Regulations</u> In certain counties, especially those with hill-farming areas, rams or uncastrated ram lambs are prohibited from certain areas at certain times of the year (Control of Rams Regulations, 1952). Increased leisure activity <u>may</u> eventually mean that bull calves will not be allowed in certain areas, e.g. fields with public footpaths.

(c) <u>The Present Situation</u>

Taking into account productivity and marketing factors, the present situation would seem to be as follows but this could change rapidly:-

1. <u>Bull calves</u> reared intensively to fatten at 12-15 months do better if not castrated but management conditions should be adequate to deal with large numbers of energetic young bulls. Codes of Practice have been drawn up. The sexes should be segregated to avoid excitement and pregnancy. In calves reared extensively, problems of management of entires may be so great as to make it more economic to castrate the bull calves.

2. <u>Boar pigs</u> destined for pork certainly need not be castrated. Uncastrated pigs reared for bacon may have developed taint at the time of slaughter. However, only a small proportion of people can reliably detect this taint (see Robertson, 1966) - the situation is in a state of flux.

3. <u>Ram lambs</u> which are going to be fattened rapidly certainly do not need to be castrated and there is doubt as to whether castration is necessary at all, except in order to obtain the guarantee price (see Fatstock Guarantee Schemes, above). Some authorities state that castration of slow-growing lambs may stunt growth to a level where the lamb does not reach a satisfactory slaughter weight in a reasonable time. However, some farmers have reported to the author a spurt in growth following cleanly done open castrations.

(d) Age at Castration

The younger the animal is castrated, the less likelihood there is of a set-back in growth but, conversely, the fatter the final carcass tends to be. Delaying castration in farm animals until beyond the age at which it is normally practised has, therefore, been suggested as a way of taking advantage of the greater growth efficiency of the entire without suffering the penalties imposed at or before marketing.

From the point of view of restraint, the easiest time to castrate pigs is before 3 weeks of age, and this is also an easy time surgically. In lambs it is easier to castrate in the first 12 hours of life, as later on the lamb becomes difficult to catch. (However, castrating lambs before 12 hours of age may be contraindicated in flocks where certain disease risks exist, e.g. watery mouth. Such lambs are usually housed and, therefore, still easy to catch at 24 - 36 hours). In calves the easiest time to castrate is at about 6-10 weeks of age - at an earlier age the testis is more friable and the technique of open castration is, therefore, more difficult.

II RESTRAINT AND ANAESTHESIA

In the U.K., the Animal Anaesthetics Act of 1964 requires that calves and lambs over the age of 3 months, and pigs and goats over the age of two months must be provided with anaesthesia or analgesia when they are castrated. (Methods are discussed in Appendix II). The Veterinary Surgeons Act (1966) as amended by Schedule 3 (Amendment) Order 1982 states that only qualified veterinary personnel may castrate animals beyond these ages.

III SURGICAL TECHNIQUES

(a) Rubber Rings

The principle of this method is that a tightly fitting rubber ring is applied to the neck of the scrotum, so that the tissues distal to it, i.e. scrotum and contents, undergo ischaemia and necrosis and eventually drop off. The method is only suitable in species in which there is a marked neck to the scrotal sac, viz. cattle and sheep. The Elastrator applicator with its rubber rings was first developed in New Zealand but it has found wide acceptance.

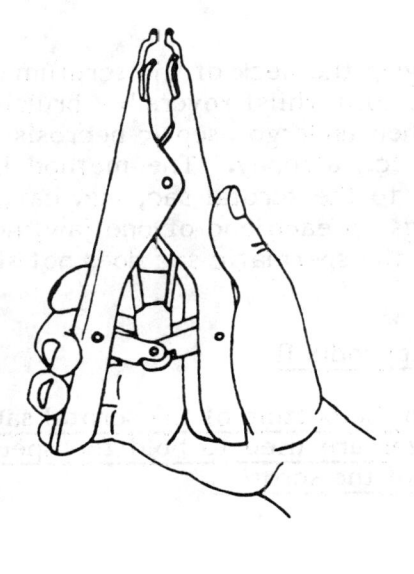

Fig. 1:8 Showing elastrator device for applying rubber rings for castration and docking

In the U.K. rubber rings may only legally be used in the first week of life. The animal is restrained by an assistant, as described previously, and the Elastrator is loaded with a rubber ring (Fig. 1:10) making sure that the rubber is not perished and that the ring is well seated on all four points. The Elastrator handles are squeezed to open the ring and the ring is moved up over the scrotum with the Elastrator points uppermost. It is important to ensure that both testes are, and remain, distal to the rubber ring. The pressure on the Elastrator handles is released so that the ring closes round the neck of the scrotum and then the Elastrator is pulled sharply distally to free it from the scrotum, leaving the ring tightly encircling the neck of the scrotum. The tissues distal to the ring undergo necrosis and eventually slough off.

Observations by Fenton, Elliott and Campbell (1958) in calves weighing about 50 Kg at the time of castration showed that the method was associated with considerably prolonged post-operative discomfort and Mullen (1964) noted that it produced a marked setback in the growth, even of young calves. Although in both sets of observations the calves were more than 1 week old, the method appears to be unsuitable for cattle.

In 4-5 week old lambs, Barrowman, Boaz and Towers (1953) noted that the method produced 5-10 minutes of discomfort but that after 30 minutes the lambs had apparently returned to normal behaviour patterns. They also noted that 70% of lambs showed some sepsis at the site, although in no case did this sepsis progress. The scrotum and contents had sloughed off in all lambs within 6 weeks. No obvious setback in growth was recorded.

The most common complication, especially in inexperienced hands, is failure to get both testes below the rubber ring and the result is induced inguinal cryptorchidism. This may not become apparent until the animal reaches puberty, when it will begin to show male behaviour. The retained testis is usually quite large, i.e. approaching normal size, and is best removed by an open castration technique. A rare complication is to include the penis in the ring and the result of this is urinary retention with eventual rupture of the urethra or bladder.

The advantage of the method is its simplicity. Its disadvantages are that it can only be done legally in very young calves and lambs and that in calves it is known to produce a marked initial setback. The method does not preclude the risk of tetanus and, as noted above, some sepsis is a common sequel. The author recommends the method only for new born lambs.

(b) Burdizzo or Bloodless Castrator

The Burdizzo is an instrument which is designed to grip the neck of the scrotum and to crush irreversibly the blood vessels of the spermatic cord whilst reversibly bruising the skin caught in its jaws. The testis and epididymis then undergo aseptic necrosis within an intact scrotal sac which shrinks as the testes, etc., atrophy. The method is only suitable in species in which there is a marked neck to the scrotal sac, viz. cattle and sheep. The instrument chosen should be one with lugs on each end of one jaw, not one without any lugs at all. The lugs help to ensure that the spermatic sac does not slip out of the crushing area.

1. The calf or lamb is restrained as described in Appendix II.

2. The right testis is grasped and drawn down into the bottom of the scrotal sac with the left hand, and the left thumb and forefinger are used to hold the spermatic sac firmly against the lateral edge of the neck of the scrotum.

3. **The right hand now holds the burdizzo and clamps its jaws across the width of the right spermatic sac only, in order to crush the right cord about 3-4 cm above the testis.**
This crush line must <u>not</u> extend across the neck of the scrotal sac, it should be at right angles to the long axis of the scrotum and it is absolutely imperative that the spermatic sac is held firmly within the jaws whilst they are clamped. The jaws should remain clamped for ½-1 minute and, after releasing the pressure, the operator should check that the cord has been crushed.

4. **The procedure is repeated about 1 cm distally.**
This double crushing is necessary to achieve satisfactory results and the second crush is made distal to the first to avoid causing unnecessary pain.

5. **The double crushing is then repeated on the left scrotal neck making sure that there is a gap of at least 1 cm (0.5 cm in young lambs) between the crush lines on the left and those on the right.** (Fig. 1:9)

If this much scrotal skin is not left intact, then the scrotal skin will undergo ischaemic necrosis, leaving an open wound which will become infected and take several weeks to heal. In the larger calf, it is not usually possible to close the burdizzo single-handed. In such an instance, an assistant should be emplyed to chose the instrument whilst the operator ensures that the spermatic sac is held laterally.

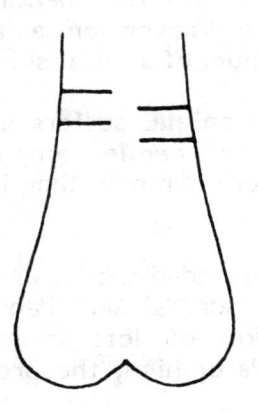

Fig. 1:9

Showing appearance of four crush marks on the skin of the scrotal neck after use of a Burdizzo.

There is a widespread belief that the burdizzo should sever the sac within the scrotal skin and some operators, therefore, try to check that the sac has been severed by testing for independant movement of the two ends. However, the instrument is not design to cut but to crush. Moreover, when testing for cutting afterwards often all that one can tell has been cut is the ductus deferens and the blood vessels may remain intact.

Some scrotal swelling develops during the days after castration and this subsides only slowly. Fenton <u>et al</u> (1958) observed that the testes remained painful for about two weeks after the operation.

The animals should be inspected by the owner 8 weeks later to check whether or not the testes have undergone atrophy. When the operation has been carried out successfully, the scrotum will feel as if it has two hard nuts inside it.

The <u>complications</u> which may follow the use of a burdizzo are as follows:-

1. The burdizzo may fail to achieve crushing of the cord. In such a case the testis may not undergo atrophy at all, or the testis may apparently atrophy but the calf or lamb later develop the characteristics of the entire male. Only a relatively small number of intestitial cells is needed to produce significant quantities of testosterone so, on palpation of the scrotum, the animal may appear to have been successfully castrated. If such an animal is not to be rejected as 'excessively male' at slaughter, the offending tissue must be removed surgically - it is usually considerably fibrosed and removal is not easy.

    The commonest cause of failure is probably a poorly maintained instrument. Bloodless castrators should <u>not</u> be stored with the jaws closed as this induces metal fatigue and results in poor performance. The efficiency of the instrument may be tested by trying it out on a piece of non-flattened straw between a folded sheet of paper. The jaws should cut the straw without breaking the paper. Should the instrument fail on three successive occasions it should be returned to the manufacturer for re-setting.

2. The operator may crush too much scrotal skin and necrosis of the skin and of all underlying tissue ensue. This is an unpleasant complication and usually results in an animal which is systemically ill and which suffers a considerable setback in growth. Treatment must be aimed at controlling the systemic illness and at cleaning and drying up the scrotum - recovery is by second intension healing with granulation and is, therefore, primarily a question of time. Prevention, as always, is better than cure and it is imperative to crush the minimum of scrotal skin.

3. The burdizzo may accidentally clamp the penis and the animal suffers urethral rupture and all its consequences. Treatment is by penile amputation. Convalescence is prolonged. This complication may be more common than is often suspected.

The <u>advantage</u> of the burdizzo is that, as no open wound is created there is, in theory, no risk of post-operative sepsis. (In practice, necrosis of the scrotal skin can occur, even when an apparently adequate amount of scrotal skin of left intact). Its <u>disadvantages</u> are its potential failure rate and, in the author's opinion, the prolonged period of pain which it produces.

The author does not recommend its use in either cattle or sheep, because he believes that the open method (see below) is more humane and can be carried out equally safely.

(c) <u>Open Castration</u>

In this method an incision is made through the layers of the scrotal wall into the lumen of the vaginal process so that the testis and adnea are exteriorized. The testis is then removed and the scrotal incision is left to heal by second intention. A whole variety of slightly different techniques have been described for this procedure, many of which undoubtedly achieve satisfactory results. The principles which it is necessary to follow to avoid post-operative complications which result in sepsis and a marked setback in growth are as follows:-

(i) the animal should be fit and clean and should be housed in a clean environment

(ii) the surgeon should be as surgically clean as possible (see especially 2 below)

(iii) the incision in the scrotum must involve the bottom of the scrotum (see especially 6 below)

(iv) only tissues which are to be removed should be touched (see especially 8 below)

The technique described below has been taught and used at the Liverpool Veterinary School for many years with a high degree of success. It is applicable to all ages and to all species of farm animals.

Different methods of restraint and anaesthesia are applicable to the different ages and species of animal as described previously. The method of haemostasis depends upon the age of the animal and is discussed below. This section details the procedure for a 6-10 week old calf and then considers modifications to it for other ages and species.

(i) **The 6-10 week old calf**

The calf should be fit and well, be out at grass or housed on fresh clean straw the day before surgery is to be done.

1. A new scalpel blade on a detachable handle is placed in a bucket, or tray with a suitable, but small quantity of antiseptic solution.
Scalpels with detachable blades are superior to others as the blade is sharp and, therefore, inflicts little pain when the skin is incised. A clear non-frothing antiseptic solution (such as a 1 in 200 solution of povidone-iodine 'Pevidine Antiseptic Solution', Berck Pharmaceuticals) is superior to any other as it allows the scalpel to be seen. A pair of sterile large haemostats should also be available.

2. The calf if caught and restrained by an assistant so that it is completely still with its scrotum accessible to the operator.
The operator must not involve himself in this procedure and will then be able to approach the task with hands which are not contaminated. In any case, he should still wash his hands thoroughly before beginning the operation.

3. The scrotum should be swabbed with antiseptic.
The swab should not be returned to the antiseptic solution containing the scalpel as the swab may obscure the latter.

4. The scrotal neck is grasped with the left hand.
The wrist should be kept out of the way of the scalpel and the testes should be well tensed into the scrotum by pressure from the fingers. If tension is inadequate it will be difficult to make a satisfactory skin incision.

5. The scalpel is taken from the antiseptic solution.
Care is taken to avoid picking up the scalpel by the blade.

6. <u>A J-shaped incision is now made through the scrotal skin, dartos into the lumen of the vaginal process, whereupon the testis should protrude freely.</u>
The incision goes vertically down the caudal ventral half of the scrotum and then continues forwards across the base of the scrotum. The incision <u>must</u> include the most ventral part of the scrotum so as to allow efficient drainage from the wound - if it involves only the caudal surface of the scrotum, a pocket is formed at the base of the scrotum from which drainage is impossible and serious post-operative swelling may result.

7. <u>The scalpel is returned to the antiseptic solution.</u>

8. <u>The testis is now firmly grasped by the finger and thumb of the right hand.</u>
There is a temptation, particularly amongst the inexperienced, to fiddle with the exposed tissue but this temptation must be resisted as this increases the risk of infection being introduced.

9. <u>Steady traction is applied ventrally and caudally in relation to the scrotum until the cord snaps.</u>
The traction should be steady, not sudden, as this stretches the spermatic cord which then snaps and undergoes considerable elastic recoil, contributing significantly to haemostasis. Should the cord snap too soon, haemorrhage seems more likely to occur.

With calves of this age, the epididymal ligament complex usually snaps readily under steady traction. The only complication is that a length of deferent duct may hang out through the scrotal incision - in such a case, the duct should be grasped and pulled out with the left hand and scaped with scalpel blade at the level of the scrotal incision until it snaps, when it will retract within the scrotal incision.

10. <u>The procedure is repeated on the other testis.</u>

(ii) **The younger calf**

Problems with the younger calf are primarily due to the smaller size and fragility of the testes. There is a tendency for the testis parenchyma to shell out of the tunica albuginea if the latter has been incised (as it often has been) and this makes it difficult to grasp the testis. The most efficient technique is to use haemostats to grasp the testis. Another possible complication is that the testis may retract if the grasp on the scrotal neck is released. With the exercise of care, the technique involved in castrating young calves is, however, readily mastered.

(iii) **The older calf**

As the calf grows beyond about 10 weeks of age, the epididymal ligament complex becomes more resistant to snapping by traction. Two solutions to this problem exist. <u>Firstly</u>, the index finger of the right hand may be used to push through the complex from cranially to caudally. This works quite well in the younger calf but is difficult in the older animal and it has the disadvantage that it involves touching the tissues which are not removed from the animal. A <u>second</u> solution is to keep hold of the scalpel handle with the right hand, to transfer the left hand from its grip on the scrotal neck to the testis which is pulled clear of the incision and then the scalpel is used to section the ligament where it attaches the tail of the epididymis to the vaginal tunic. This method is applicable to all ages of calves and does not involve the operator's hands in touching any structure going back into the animal.

Traction will provide adequate haemostasis for calves up to 4 or 5 months of age. From then up to about 10 months of age, haemostasis is best achieved by twisting the vascular portion of the cord only. The tunic must, therefore, be stripped up into scrotum. This can be done by grasping it with a large pair of haemostats and pushing the tunic away into the scrotum. These same haemostats may then be clamped on the vascular portion of the cord almost along the line of the cord, not transversely. An index finger is then inserted through one of the finger holes and the cord twisted round and round whilst pulling distally. Alternatively, the testicle itself is grasped and twisted. The twisted cord snaps and retracts after about 20 full-circle twists. The operator must ensure that no hair gets caught in the twisted cord. Should the animal struggle during this procedure, the tension on the cord should be relaxed not tightened, otherwise it may snap prematurely and haemorrhage may occur.

In the older bull calf and mature bull, it is better to use an emasculator for haemostasis (p.12). Some authorities prefer to use ligature but to do so safely requires a high degree of surgical cleanliness and has little advantage, if any, over the use of an efficient well designed emasculator (p.15).

(iv) The lamb

In small lambs, it is recommended that the left hand be used to pull the bottom of the scrotum, but not the testes, distally and the scalpel be used to cut off the bottom of the scrotum whereupon the testes drop out through the incision and may be separated by straight-forward traction. This method has the advantage that it avoids any possibility of incising the testis and, therefore, avoids the testis parenchyma shelling out of the tunica albuginea and so making it difficult to grasp the testis. Otherwise, the only difference between castrating lambs and calves is in the method of restraint. Contrary to published opinion, open castrations can safely be done in adult sheep, provided the clostridial vaccination status is satisfactory and the standard of cleanliness is high.

(v) The piglet

The golden rule of castrating piglets is to ensure that the sow is securely restrained whilst the piglets are being handled. The sub-anal position of the scrotum makes the piglet susceptible to post-castration infection arising from the faeces, so scouring piglets should never be castrated.

The sub-anal position of the testis and virtual absence of a scrotum with a definitive neck in the pig also means that a slightly different technique has to be used to tense the testis into the scrotum. The middle finger of the left hand is placed in the inguinal region to push and the thumb and first finger used to hold the testis correctly aligned in the scrotum. It is important to hold the testis in the scrotum accurately and to make the incision in the middle of the scrotum as, especially in very young piglets, there is a risk of incising the muscles of the thigh.

In the adult boar, the post-operative swelling that develops may be considerably reduced by suturing the scrotal skin. The extra care with surgical cleanliness and the extra time involved in suturing are repaid by a more rapid convalescence.

(d) <u>Other Techniques</u>

In the book of Leviticus in the Bible, it is recorded at Chapter 22 and Verse 24 that a male animal may not be offered as a sacrifice if its testicles have been bruised, twisted, torn or cut. The references to tearing and cutting presumably refer to the techniques described under (c) above. The reference to bruising is to the barbaric practice of destroying the substance of the testes by crushing the testes between two rocks. This technique was certainly still in use in the last century and is probably still in use in underdeveloped countries today. It is, however, not to be recommended! The reference to torsion may not be to twisting the cord, as described above on p.29 but to a technique which involves twisting the neck of the spermatic sac percutaneously, a technique that sounds rather difficult to perform.

(e) <u>Complications</u>

The complications which may follow an 'open' castration in farm animals are much the same as those described for the horse (section p.16). <u>Haemorrhage</u> is less common in farm animals, primarily because they are castrated at a much younger age. <u>Eventration</u> is extremely rare in cattle and sheep but may occur in the pig if an inguinal hernia was not noticed prior to castration. <u>Post-operative oedema</u> is less of a problem in the farm animals than it is in horses, but fluid accumulation, non-specific infection and abscesses may all produce gross scrotal swelling and treatment as outlined for the horse is appropriate in all species. In sheep in particular there is a danger of clostridial infection supervening so the vaccination or immune status of the animal needs confirming.

Scrotal abscesses and scirrhous cord may both follow castration of the pig. In the former case surgical excision is preferable to provision of drainage as the latter is difficult to achieve in the pig.

An unusual and rare complication of castration by traction is '<u>gut-tie</u>', sometimes also called pelvic hernia. The bowel becomes incarcerated and may eventually strangulate in one of two ways. Traction on the spermatic cord may tear the peritoneal fold of the deferent duct which fixes the duct to the pelvic wall and loops of bowel may pass through this hiatus and become incarcerated. Alternatively the deferent duct may retract into the abdomen after castration, become adherent to the abdominal wall or to viscera and thus form a loop in which intestine may become incarcerated. The incarceration/strangulation may occur many weeks or months after castration and is almost invariably on the right side as the presence of the rumen nearly precludes its occurring on the left. The clinical signs which develop are those of intestinal obstruction, viz. colic, absence of faeces, malaise, fast pulse. Rectal examination will reveal the presence of trapped bowel usually just cranial to the pelvic inlet. The tense band of the deferent duct may also be palpable. The condition is amenable to surgical correction through a right flank laparotomy and the prognosis is good if trangulation has not supervened.

A review of the literature and a case were presented by O'Connor (1972).

IV   ALTERNATIVES TO CASTRATION

A number of techniques of partial castration or of sterilisation have been described. Although advantages over conventional techniques have been claimed for them, these advantages have not been completely vindicated and the methods are not widely used.

In the author's opinion there is no doubt that most animals 'castrated' in these ways do not lose all of the interstitial cells of the testis and that those cells which remain are able to produce androgens in considerable quantities. Therefore, apart from being infertile, these animals probably have little advantage over entires in terms of growth performance or behaviour.

(a)   Baiburtcjan's Method

This was described in 1961 by Baiburtcjan in Russia. It involved making a stab incision through the lateral scrotal skin into the testicular parenchyma and carefully squeezing out the substance of the testis through the incision.

(b)   Epididymectomy

Some Australians have described a technique involving removal from lambs of the epididymal tails and, in some cases, removal of one testis, or even one and a half testes. This technique is also discussed in Chapter 6.

(c)   Induced Cryptorchidism

A technique for inducing cryptorchidism by deliberately placing Elastrator rings (p.23) below the testes and so eliminating the scrotum and retaining the testes in the inguinal region has also been described.

D. MISCELLANEOUS DISORDERS REQUIRING CASTRATION

This section is concerned not with disorders of spermatogenesis but with grosser diseases of the scrotum and testis.

I TUMOURS OF THE TESTIS

(a) <u>Teratoma</u>

This is a tumour which contains tissue derived from more than one of the three primary germ layers, viz. ectoderm (often neurectoderm), mesoderm and endoderm. It is virtually unknown in domesticated animals other than the horse. Teratomas occur more commonly in the cryptorchid testis but they can occur in scrotal testes. They appear to be more common in the heavy draught breeds and possibly in arabs than in other breeds. The tumours are almost certainly benign and are probably well established at an early age. Sometimes they are grossly cystic, whilst at other times they are more solid with lumps of cartilage or bone. They often contain either teeth or hair (see Cotchin, 1978).

Generally speaking, they are symptomless and only arise as incidental findings at castration.

(b) <u>Other Tumours</u>

Other testicular neoplasms include interstitial cell tumours (which have been incriminated as a cause of viciousness in horses), seminomas (which may be malignant or benign), Sertoli cell tumours, haemangiomas, lipomas and adenocarcinomas. The Sertoli cell tumours may be endocrinologically active, producing signs of feminisation. Otherwise, increase in testicular size is the predominant presenting sign. Castration is the only possible treatment. However, other causes of testicular enlargement should be eliminated by careful clinical inspection or by biopsy if necessary. Such causes include haematoma, orchitis and hypertrophy.

II VARICOCELE

This is an uncommon condition although it may occur in the horse and ox. The veins of the spermatic cord become varicose and the cord assumes the nature of a soft knotty elongated swelling of varying size. As a rule it causes no inconvenience. Occasionally, however, the enlarged cord may become strangulated in the inguinal canal (Rutherford, 1891) and produce signs almost indistinguishable from a strangulated inguinal hernia (see p.62.) - indeed, the two conditions may only be differentiated at surgery.

The condition may present itself in the course of a routine castration in which case especial care with haemostasis during castration is all that is necessary. If the condition is in the strangulated state, immediate surgery is indicated involving castration with a secure method of haemostasis. In either case, the author would employ a ligature of 5 metric synthetic absorbable suture material (doubled) on the blood vessels.

III HYDROCELE

In this condition there is an excess of peritoneal fluid in the lumen of the vaginal process. Whilst it may accompany ascites and the older literature attributes it to a chronic vaginalitis following injury to the scrotum, the cases the author has seen have all been congenital and in shire horses (see Wright, 1963). If blood should be mixed with the fluid it is called an haematocele. Hydrocele often accompanied tuberculous orchitis in cattle, but as tuberculosis is now uncommon, hydrocele from this cause is rare.

It is characterised by a soft, fluctuating scrotal swelling, sometimes bilateral. Palpation of the scrotum and inguinal area fails to reveal the presence of intestine or omentum although, in the author's experience in shires, the inguinal canal is larger than normal. The presence of fluid can make it difficult to palpate the testis, especially in young animals in which the testes are small.

The condition does not usually affect the animal except that pressure from the fluid may cause atrophy of the testis.

Diagnosis is usually easy on clinical grounds but may be confirmed by exploratory puncture under surgically clean conditions. If the case is one of hydrocele a serous, amber-coloured fluid escapes. The condition has a superficial resemblance to inguinal hernia (see p.59FF) and to varicocele (see p.32) and must be distinguished from cystic ends to cords (see p.20) in the gelding.

Medical treatment is unlikely to be of lasting benefit. If the animal is to be castrated this should be done by a 'closed' method (p.19) as the inguinal canal is probably larger than normal.

IV TRAUMA AND ORCHITIS

(a) Aetiology

Damage to the scrotum and testis may be caused in a variety of ways. Examples include lacerations of the scrotum in horses jumped over hedges and dermatitis of the scrotum in the stallion (e.g. as a result of a blister applied to the hocks being transferred to the scrotum by the tail), frostbite as a result of standing in deep snow in the bull, mange in the ram, and fighting between males in the pig.

Glanders and dourine may all cause orchitis in the stallion, as also may migrating strongyle larvae. In the bull orchitis may be caused by Br. abortus, C. pyogenes, Actinomyces bovis and Mycobacterium bovis and in the boar orchitis may be caused by Br. suis, although this organism is not present in the U.K. in all cases, orchitis may be accompanied by epididymitis, vesiculitis, ampullitis and vaginalitis.

(b) Pathogenesis and Clinical Signs

A varying degree of inflammation develops following any injury, and is accompanied by oedema and clinically apparent as swelling.

If a wound penetrates to the lumen of the vaginal process or a testicular abscess bursts through tunica albuginea into the vaginal process, then peritonitis may ensue. Otherwise, the inflammatory process may result in adhesions within the vaginal process or the accumulation of an exudate (see also Hydrocele, above).

The swelling and/or pain in the testicle may cause the animal to move with the hind limb of the affected side abducted. In severe cases the animal may be reluctant to move at all and systemic illness may be apparent. Reluctance to serve is usual.

Initially in orchitis the testis may be flabby but later it becomes firm with fibrosis and possibly calcification. If the epididymis becomes affected, this is usually most clinically apparent in the tail which is initially swollen, hot and painful, eventually becoming hard, fibrous, shrunken and irregular. Vesiculitis and ampullitis can be detected on rectal examination.

(c) Prognosis and Treatment

Even though an injury may be limited to the scrotum, or disease processes to one testical, infertility usually results. In the ipsilateral testis, infertility may be due to increased tension inside the tough capsule interfering with normal blood supply. Infertility in the contralateral testis may be due to local inflammatory heat, toxins and allergic reactions. Such infertility may be temporary or permanent, probably depending on the length of time for which the spermatogenic epithelium is insulated.

It is important, therefore, to limit any inflammatory process. Clean scrotal wounds may be sutured, but infected ones are best dealt with by immediately castrating the animal on the affected side. Unilateral orchitis is also best dealt with by removal of the affected organ.

Although castration may seem a drastic approach, it does remove the seat of the trouble and allow rapid resolution. It is important to realise that it takes about 6-8 weeks for sperm to arrive in semen from the beginning of spermatogenesis. Therefore, at least this interval should be allowed to elapse between the disappearance of all signs of inflammation and rejection of the animal as sterile - degeneration can occur quickly, regeneration may be slow.

V   TORSION OF THE SPERMATIC CORD

Torsion of the spermatic cord of the horse was first described by Rutherford (1891), and later by Horney and Baker (1975) and by Pascoe, Ellenburg, Culbertson and Meagher (1981). It is a condition of the entire male horse and is also documented in rams, pigs and dogs.

The horse presents clinically as an acute, mild colic. The skin of the scrotum on the affected side is cold and moist and testicle and spermatic cord are palpably enlarged and may be painful. Retraction of the testis may or may not occur. Rectal exploration will reveal slightly enlarged, normal structures passing through the deep ring. The condition should be suspected in any stallion with colic and requires differentiation from varicocoele (p.32) and strangulated inguinal hernia (p.62) or rupture (p.64).

Surgical exploration under general anaesthesia is indicated and will reveal a distended spermatic sac which when opened will contain an engorged, dark red testicle. In Pascoe's case the epididymal tail was attached to the testis by an elongated proper ligament and this arrangement appeared to have allowed the testis to rotate through 360°). The only appropriate treatment is castration on he affected side.

360° torsion should be distinguished from a 180° torsion in which the epididymal tails lie cranial to the testis - in the author's experience this latter finding is a rare anatomical abnormality which does not affect the animal's fertility and is not a therapeutic indication for castration.

E. REFERENCES

Arthur, G.H. (1956) A Study of some aspects of descent of the testes and cryptorchidism in domestic animals. Fellowship Thesis for the Royal College of Veterinary Surgeons, London.

Backhouse, K.M. and Butler, H. (1960) The gubernaculum testis of the pig (Sus Scrofa). Journal of Anatomy 94 : 107-120.

Baiburtcjan, A.A. (1961) "A new method of increasing livestock productivity (castration with hormonal integrity)". Translated from Russian by G.R. Ritchie (1963) in Animal Breeding Abstracts 31 : 1-21.

Barber, S.M. (1985) Castration of horses with primary closure and scrotal ablation. Veterinary Surgery 14: 2-6.

Barrowman, J.R., Boaz, T.G. and Towers, K.G. (1953) Castration of lambs: comparison of the rubber-ring ligature and crushing techniques. Empire Journal of Experimental Agriculture 21 : 193-203.

Baumans, V., Dijkstrom, G. and Wensing, C.J.G. (1983) The role of a none androgenic factor in the process of testicular descent in the dog. International Journal of Andrology 6 : 541-552.

Baumans, V., Dieleman, S.J. Wouterse, H.S., Van Tol, L., Dijkstrom, G. and Wensing, C.J.G. (1985 Testosterone secretion during gubernacular development and testicular descent in the dog. Journal of Reproduction and Fertility. 73 : 21-25.

Bergin, W.G., Gier, H.T., Marion, G.B. and Coffman, J.R. (1970) A developmental concept of equine cryptorchidism. Biology of Reproduction 3 : 82-92.

Cotchin, E. (1978) Equine testicular teratoma in Tumours of Early Life in Man and Animals. Ed. L. Severi, VIth Perugia Quadrennial International Conference on Cancer, Perugia, Italy.

Cox, J.E. (1984) Castration of horses and donkeys with first intention healing. Veterinary Record 115 : 372-375.

Degive, A. (1875) De la castration des animaux cryptorchides. Annuales Medicine Veterinaire 11 : 629-647 and 693-720.

Fenton, B.K., Elliott, J. and Campbell, R.C. (1958) The effect of different castration methods on the growth and well-being of calves. Veterinary Record 70 : 101-102.

Formston, C. (1951) Equine castration. Veterinary Record 63 : 18-20.

Gier, H.T. and Marion, G.B. (1969) Development of mammalian testes and genital ducts. Biology of Reproduction 1 : 1-23.

Horney, F.D. and Baker, C.A.V. (1975) Torsion of the testicle in a standard bred. Canadian Veterinary Journal 16 : 272-273.

Kiley, M. (1976) A review of the advantages and disadvantages of castrating farm livestock with particular reference to behavioural effects. British Veterinary Journal 132 : 323-331.

Lowe, J.E. and Dougherty, R. (1972) Castration of horses and ponies by a primary closure method. Journal of the American Veterinary Medical Association 160 : 183-185.

Mullen, P.A. (1964) Some observations on the effects of the method and time of castration for Barley Beef Production. British Veterinary Journal 120 : 518.

O'Connor, J.P. (1972) A case of pelvic hernia or 'gut-tie' in a Bullock. Irish Veterinary Journal 26 : 251-252.

Palmer, S.E. (1984) Castration of the horse using a primary closure technique. Proceedings of the Annual Connection of the American Association of Equine Practitioners 30 : 17-20.

Pascoe, J.R., Ellenburg, T.V., Cutherbertson, M.R. and Meagher, D.M. (1981) Torsion of the spermatic cord in a horse. Journal of the American Veterinary Medical Association 178 : 242-245.

Roberts, E.J. (1964) Some modern surgical operations applicable to the horse. Veterinary Record 76 : 137-142.

Robertson, I.S. (1966) Castration in farm animals: its advantages and disadvantages. Veterinary Record 78 : 130-140.

Rutgers, L.J.E. and Merkens, H.W. (1985) Primaire sluiting van de scrotaalwond bij de castratie van de hengst. Vlaams Dier geneeskunde Tijdschrift 108 : 717-722.

Rutherford, R. (1891) Scrotal hernia. Veterinary Record 4 : 194-195.

Vaughan, J.T. (1978) Urogenital Surgery. Proceedings of the Annual Convention of the American Association of Equine Practitioners 24 : 513-566.

Wensing, C.J.G. (1968) Testicular descent in some domestic mammals. I. Anatomical aspects of testicular descent. Proc. Kon. Ned. Akad. Wetensch. C71 : 423-434.

Wensing, C.J.G. (1973a) Testicular descent in some domestic mammals. II The nature of the gubernacular change during the process of testicular descent in the pig. Proc. Kon. Ned. Akad. Wetensch. C76 : 190-195.

Wensing, C.J.G. (1973b) Testicular descent in some domestic mammals. III. Search for the factors that regulate the gubernacular reaction. Proc. Kon. Ned. Akad. Wetensch. C76 : 196-202.

Wensing, C.J.G. (1973c) Abnormalities of testicular descent. Proc. Kon. Ned. Akad. Wetensch. C76 : 373-381.

Wensing, C.J.G. and Colenbrander, B. (1973) Cryptorchidism and inguinal hernia. Proc. Kon. Ned. Akad. Wetensch. C76 : 489-494.

Wright, J.G. (1963) The Surgery of the inguinal canal in animals. Veterinary Record 75 : 1352-1363.

Chapter Two
CRYPTORCHIDISM

A. THE PHENOMENON

I   DEFINITIONS

<u>Cryptorchid</u> = animal with one or two testes retained somewhere along the normal pathway of descent.

<u>Ectopic Testis</u> = testis which has <u>deviated</u> from the normal pathway of descent. These may lie alongside the penis cranial to the scrotum.

<u>Monorchid</u> = animal with only one testis, the second never having developed. It does <u>not</u> mean an animal with only one scrotal testis. An animal with one testis scrotal and one retained is a unilateral cryptorchid. Monorchidism is rare in all species.

<u>Anorchid</u> = animal in which neither testis ever developed. It does <u>not</u> mean an animal with no scrotal testes. An animal with two testes retained is a bilateral cryptorchid. Anorchidism is extremely rare in all species.

<u>Polyorchid</u> = animal with more than two testes. Such animals have been described (e.g. Foster, 1952). Certain diagnosis, however, requires the finding of three testes at one and the same time. Otherwise the possibility of wrong identification cannot be ruled out. Other possible explanations of apparent polyorchidism exist. The most likely is that the epididymis and the testicle have been mistaken for separate glands (this is most likely to happen if the animal is an incomplete abdominal cryptorchid - see p.40 below. Other possible explanations of polyorchidism are that a hard cyst (see p.20) on the cord has been diagnosed as an extra testicle.

<u>Rig</u> = Laymen's term for cryptorchid or horse which behaves like one, i.e. has no visible testes but shows masculine behaviour. 'Rig' may derive from ridgeling (perhaps because the testis lay high up under the back, i.e. the ridge) or from the dialect word 'rig' meaning a joke or trick, or a licentious person.

<u>False rig</u> = horse which has had both testes completely removed, but is showing some form of stallion like behaviour.

II   THE RUMINANTS AND THE PIG

Cryptorchidism is generally rare in the sheep, goat and cattle but occasionally up to 5% of ram lambs in a flock may be affected (Arthur, 1975). Ectopic testes are more common than cryptorchid testes in these species.

Sporadic cases of cryptorchidism occur in most pig units but, as with sheep, occasional 'outbreaks' occur. The rig pig without a scrotal testis does not produce 'taint' and there is, therefore, no need to operate to remove the cryptorchid testis from such a pig. Nevertheless, the scrotal testis should be removed to prevent the animal breeding and the pig should then be fattened for slaughter.

Almost invariably, the retained testis and its associated epididymis are entirely within the abdomen.

III THE HORSE

It is in this species that cryptorchidism is the most significant because it poses problems of diagnosis, as well as anaesthesia and surgery. The description which follows is based primarily upon observations made by the author and his colleagues at Liverpool (see Cox, Edwards and Neal, 1979). Other surveys of large numbers of cases have been made by Stanic (1960) and Stickle and Fessler (1978).

(a) Temporary Inguinal Retention

This type of cryptorchidism occurs predominantly, but not entirely, in ponies and is characterised by small testes weighing less than 40 gm. The larger testes can generally be palpated in the inguinal region of the quiet standing horse and testes of all sizes are usually readily palpable in (and can be readily removed from) the anaesthetised horse in dorsal recumbency. If not removed, these testes grow and descend into the scrotum usually before the animal becomes a three year old.

The condition is usually unilateral and is predominantly (> 3 out of 4) right sided.

The testis has the relationships and the macroscopic appearance of a normal scrotal tetis. As the testis is smaller than most scrotal testes it appears to have a relatively large epididymis (Bishop, David and Messervey, 1964), especially in the older, larger horse.

The microscopic appearance of the testis is that of an immature testis with spermatogenic tubules occupying most of the substance of the testis and with few interstitial cells. The spermatogenic tubules contain supporting cells and spermatogonium-like germ cells which fill the tubules. However, the nearer the testis gets to the scrotum, the more mature and normal its appearance becomes - testes which are almost scrotal may produce spermatozoa in some tubules.

(b) Permanent Inguinal Retention

This type occurs in all types of horse and is characterised by testes which generally weigh more than 40 gm and may be misshapen. They cannot always be palpated readily in the standing horse and may sometimes be palpable only with difficulty in the anaesthetised horse in dorsal recumbency. The testis must, by definition, have passed through the deep inguinal ring, but it may still be retained partly within the inguinal canal. The testis usually has a short vaginal tunic and can be brought to the outside only with difficulty. The condition is usually unilateral, the left or the right testis being retained with equal frequency. Occasionally, the contralateral testis may be abdominally retained.

In the young horse, the microscopical appearance of the testis is like that for a temporarily retained testis but in the older horse, the number of interstitial cells increase and the tubules show evidence of degeneration, e.g. the supporting cells become vacuolated and the cells aggregate round the periphery of the tubule.

(c) <u>Complete Abdominal Retention</u>

In this type of abdominal retention both testis and epididymis are completely retained within the abdomen. The testis itself is suspended in the abdominal cavity from the sub-lumbar region by a fold of peritoneum reaching from just caudal to the kidney to the bladder (Fig. 2:1). The cranial edge of this fold contains the spermatic vessels which pass dorsally and cranially. The central edge is occupied in succession as one passes caudally by (i) the testis, (ii) the remains of the proximal part of the true gubernaculum joining testis to epididymal tail (= proper ligament of the testis), (iii) the epididymal tail itself and (iv) the deferent duct. The proper ligament of the testis is usually considerably elongated thus separating the epididymal tail from the tetis by quite some distance. The body of the epididymis is supported in a minor lateral fold of the peritoneal fold thus creating an epididymal sinus similar to that of the normal descended testis. Another small fold of peritoneum passes internally from the epididymal tail to the area of the deep inguinal ring. This small fold contains the remains of the distal part of the true gubernaculum (=ligament of the tail of the epididymis). In some cases a small vaginal process with a small cremaster muscle has developed in the inguinal canal and only the smaller fold of peritoneum containing the ligament of the tail of the epididymis enters the inguinal canal. There can sometimes be found in the inguinal region a few fibrous strands running distally from the tip of such a vaginal process - this is the remains of the infravaginal part of the gubernaculum. The similarity between these structures and their relationships and those of the fetal gonad (p.2 Fig. 1:1) is striking.

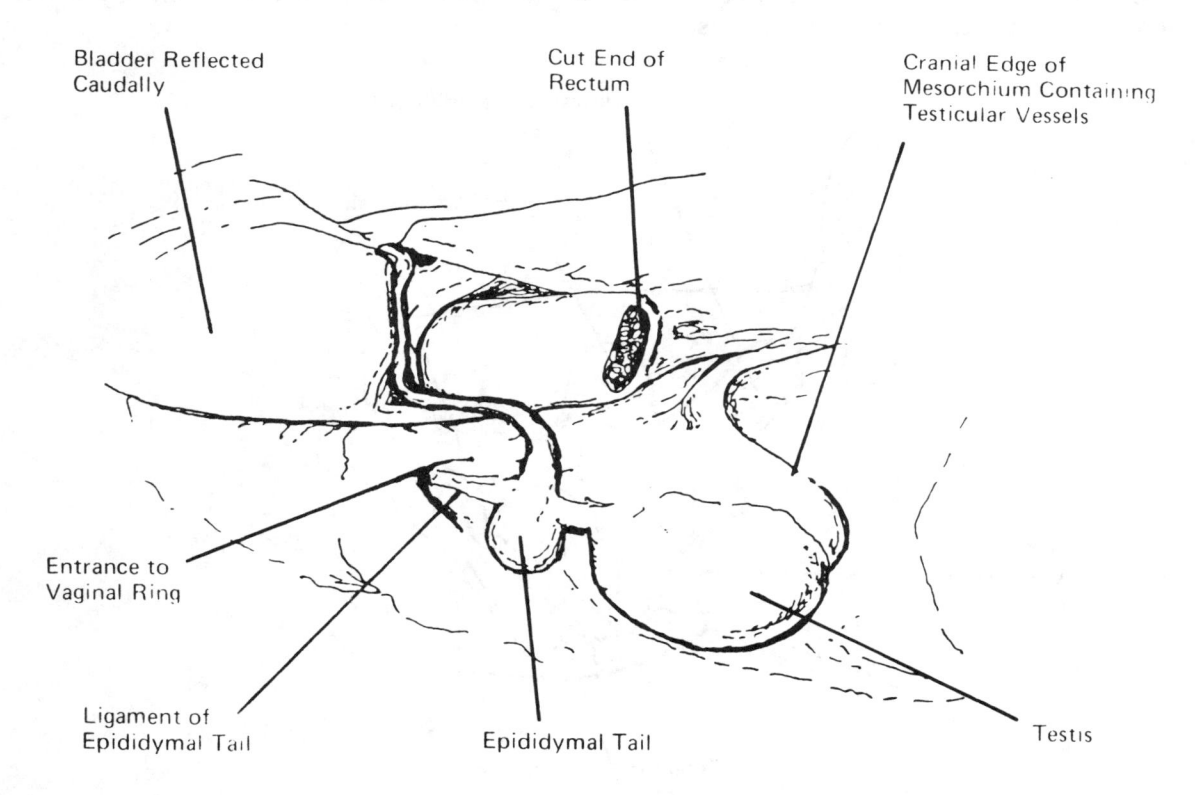

Fig. 2:1 Dissection of a cryptorchid pig to show relationship of an abdominally retained testis.

The abdominally retained testis, therefore, is relatively freely mobile within the abdomen. Although it usually lies close to the deep inguinal ring, it may become mixed up with coils of intestine, lie dorsal to the rectum or lateral to the bladder. In exceptional cases it is adherent to the wall of the abdomen or to other organs such as the spleen and may then be exceedingly difficult to identify.

The abdominally retained testis is usually small, weighing between 10 and 50 gm. It is then characteristically flabby. Occasionally, however, it may be grossly enlarged, due to cyst or teratoma formation. The testis consists of islands of spermatogenic tubules with associated interstitial cells in a sea of loose connective tissue. The spermatogenic cells rarely progress beyond primary spermatogonia although spermatozoa have been reported (Hobday, 1914). It seems that the proportion and density of the fibrous tissue increase with age.

(d) Incomplete Abdominal Retention

In this type of abdominal cryptorchid, the vaginal process is well developed. It has an attached cremaster muscle and contains the epididymal tail as well as part of the deferent duct and part of the body of the epididymis. These latter two structures each have a fold of peritoneum and pass proximally through the vaginal ring toward the bladder and testis respectively (see Fig.2:2). The length of vaginal tunic is variable, and it may not extend beyond the limits of the inguinal canal or it may reach the scrotum. The vaginal tunic and contents can sometimes be palpated in the standing horse and can often be felt in the inguinal area of the anaesthetised horse in dorsal recumbency when they may be mistaken for a small testis.

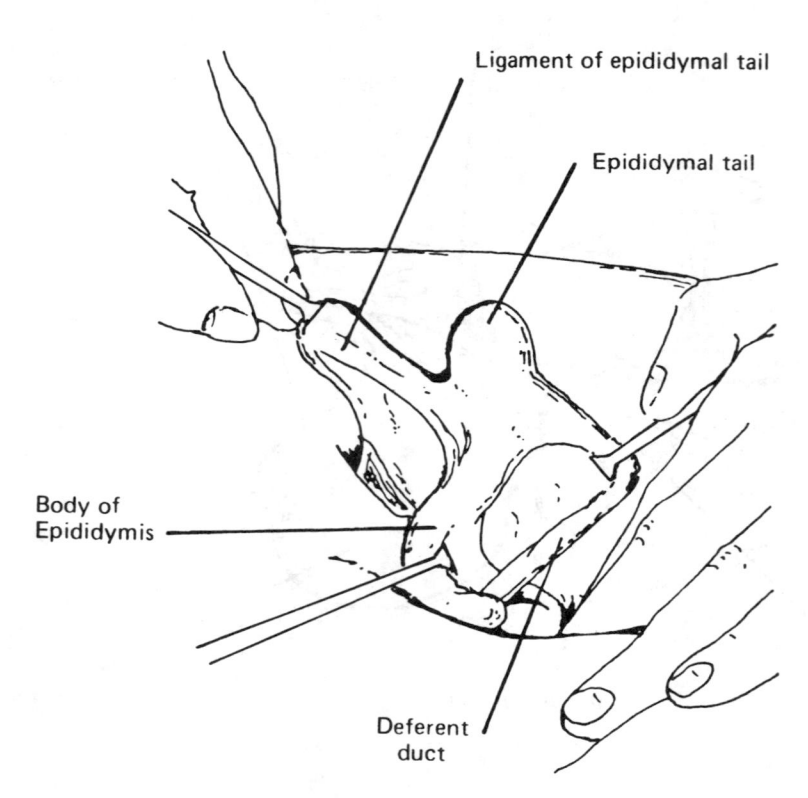

Fig. 2:2 Sketch from life showing surgical findings in a case of incomplete abdominal cryptorchidism. The vaginal tunic has been incised along its length to expose its contents.

The testis is within the abdomen but usually close to the deep inguinal ring and lacks as much potential mobility as that in a case of complete abdominal retention. Its texture and microscopic structure are similar to that of a complete abdominal cryptorchid.

Incomplete retention occurs in 50% of cases of unilateral right-sided abdominal retention but in only about 15% of unilateral left-sided cases. In cases with bilateral retention both testes are usually 'complete' or both are 'incomplete' - only two cases of bilateral absominal retention have been seen at 'Leahurst' since 1955 in which one side was 'complete' and one 'incomplete'. Irrespective of whether the epididymal tail is or is not descended, in ponies the left testis is retained as frequently as the right but in thoroughbreds, quarter-horses, trotters and larger horses (cobs, shires), left-sided cases outnumber right-sided cases by a ratio of two to one.

(e) The foal

Palpation of the scrotum of the new born foal can be extremely confusing. At this stage the gubernacular complex is large, and it and the epididymal tail may together be larger than the testis. If the epididymal tail has descended, therefore, the foal may be designated normal although, in fact the testis is actually in the abdomen. Even when the testis and epididymis both remain within the abdomen, the gubernacular complex may be so large as to be mistaken for a small testis. This has probably given rise to the mistaken concept that all foals are born with two normal scrotal testes.

IV THE CAUSES OF CRYPTORCHIDISM

It is often stated that cryptorchidism is inherited but usually without much satisfactory data about the mode of inheritance. Studies in Angora goats, summarised by Lush, Jones and Dameron (1930), suggested that in many cases at least two pairs of autosomal genes were involved but that other cases could not be explained in this way and they probably arose sporadically. Clinical observations in horses suggest that some cases are inherited and others sporadic. The genetics of cryptorchidism in pigs have been thoroughly studied by Mikami and Freedman (1979) who confirm its hereditability. The apparently "sporadic" cases may, therefore, be due to the complexity of the genetics. (see also Leipold, De Bowes, Bennett, Cox and Clen, 1986 and Hayes, 1986.)

A deficiency of androgens has often been implicated as *the* cause of cryptorchidism but the role of androgens in normal testicular descent in the domestic animals (as opposed to some small laboratory animals) is equivocal. Moreover, it is rather difficult to explain unilateral cryptorchidism on a hormonal basis.

Studies by Bergin et al (1970) on the horse emphasised physical factors in the development of cryptorchidism, especially (i) undue stretch of the true gubernaculum, (ii) failure of the whole gubernaculum to expand and so dilate the inguinal canal, (iii) failure to develop adequate intra-abdominal pressure for extension of the vaginal process into the gubernaculum and (iv) displacement of the testis into the pelvic cavity where it would be held by other organs. The work of Wensing in the pig and dog has also refocused attention on the gubernaculum, especially gubernacular development which is abnormal in site (i.e. not inside and distal to the inguinal canal) abnormal in direction (e.g. not directed scrotally) or abnormal in extent (e.g. inadequate). However, whether cryptorchidism arises from factors inherent in or external to the gubernaculum still remains unknown.

V THE SIGNIFICANCE OF CRYPTORCHIDISM

Most large male animals are destined to be castrated either so that they can be kept quietly to slaughter weight (so cattle, sheep and pigs) or so that they can be managed more easily as leisure animals (so horses and donkeys). Cryptorchidism, therefore, reduces the facility with which castration may be carried out.

The uncastrated cryptorchid poses other problems. Although the retained testis will be sterile, the normal scrotal testis of a unilateral case will produce sufficient spermatozoa for the male to be fertile - indeed it is probable that he will not be significantly less fertile than a male with two scrotal testes. Nevertheless, a unilateral cryptorchid should not be used as a stud animal as, if it has hereditary cryptorchidism, the incidence of cryptorchidism and the attendant problems will increase.

The interstitial cells of a retained testis continue to be able to produce androgens and, although they are reduced in number, it appears that they work harder - cryptorchid males with bilateral or unilateral retention have similar concentrations of testosterone in peripheral plasma to normal males (Cox, Williams, Rowe and Smith, 1973). Because therefore, androgens are still secreted by the retained testis, the animal with such a testis or testes goes on behaving as though he were entire male. There is a popular view that cryptorchid animals are more vicious than normal entires, but this is not the case.

B. APPROACH TO A CRYPTORCHID HORSE

I CLINICAL EXAMINATION

Any horse in which two testes are not visible and readily palpable in the scrotum must be suspected of being a cryptorchid. Careful examination in the standing horse (Fig. 1:6), possibly after tranquillisation, may allow a testis to be palpated which has been temporarily retracted under the influence of fear or of cold weather or hands, or which has been retained only just dorsal to the scrotum. Deep palpation of the inguinal region may allow an inguinal testis to be palpated but there are structures other than testes (e.g. remains of spermatic sac, spermatic sac containing epididymal tail) which may be present in the inguinal area and which may be difficult to distinguish from a testis by palpation alone. Moreover, not all inguinally retained testes are palpable. If, therefore, more detailed examination shows a retained testis to be present, the animal may be dealt with quickly but if a testis is not palpable, the surgeon must be prepared for anything.

It is sometimes stated (e.g. Adams, 1964; O'Connor, 1971) that rectal examination of the vaginal ring prior to operation is useful to distinguish inguinal cryptorchidism from abdominal cryptorchidism. However, it is rarely possible to palpate the testis and, therefore, the evidence on which a diagnosis must be based is indirect. Only in the complete abdominal cryptorchid will no structure which can be palpated to pass through the vaginal ring. In the incomplete abdominal cryptorchid the deferent duct and the body of epididymis pass through the ring but in cases where the testis has descended through the inguinal ring the deferent duct and spermatic vessels will be found to pass through the ring. As it is not possible readily to distinguish blood vessels from body of epididymis, it is not possible to distinguish incomplete abdominal cryptorchidism from cases in which the testis has passed the vaginal ring.

II DIFFERENTIATION FROM GELDINGS

Until the development of a blood test, diagnosis of the horse with no visible testes but displaying male behaviour has been based primarily upon clinical features and surgical investigation.

(a) Palpation

The palpation of two scars in the scrotal area is widely believed to be diagnostic of a gelding. The skin of the scrotal area is pulled between fingers and thumb and gently rolled. In an animal castrated by an 'open' method, the stump of spermatic sac will generally be felt to roll between the fingers. This method is still widely applicable in examinations prior to sale.

(b) Behaviour

It has been stated that the gelding can be distinguished from the cryptorchid by the former's inability to achieve penile erection and intromission (Wright, 1963). However, a large number of geldings are quite able to achieve erection and to serve mares.

(c) Surgical

Under general anaesthesia, the finding in each inguinal region of a stump of a spermatic sac containing deferent duct and remains of spermatic vessels proves that the animal has been castrated. However, it is not in all geldings that these structures can be found.

(d) Blood Test

The development of a simple blood test for distinguishing geldings showing male behaviour from cryptorchids was the result of work in the author's laboratory (Cox, Williams, Rowe and Smith, 1973; Cox, 1975; Cox, 1982; Cox, Redhead and Dawson, 1986).

In horses over three years of age, a single blood sample is taken for the measurement of oestrone sulphate - cryptorchids have concentrations in excess of 400 pg/ml whereas geldings have less than 100 pg/ml. In younger horses and in donkeys of all ages, two blood samples are required, one prior to and the second 30 to 120 min after the intravenous injection of 6000 i.u. of human chorionic gonadotrophin. In this instance, testosterone concentrations are measured and a diagnosis of cryptorchidism based upon concentrations in excess of 100 pg/ml or a rise in concentration in response to the hCG or, preferably, both. The sample(s) should be submitted to a laboratory accustomed to analyse samples from horses.

A discussion of the problem of the animal which the blood test shows has been castrated (false rigs) will be found in Cox (1986).

III REMOVAL OF A SINGLE TESTIS

There is a widespread belief that removal of the scrotal testis from a unilateral cryptorchid promotes descent of the undescended testis. This is not only fallacious belief but is potentially legally dangerous.

There is no evidence that removal of the scrotal testis actually <u>promotes</u> descent of an undescended testis. If the retained testis is a temporarily retained inguinal one, it will descend spontaneously. If it is a permanently retained inguinal testis it will not descend further. If the retained testis is abdominal it will never descend. Nothing will be gained by waiting and, therefore, the veterinary surgeon who is asked to castrate a horse should never remove a single testis but should either be prepared to remove both testes or to refer it to someone who will.

Should the owner ask for a single scrotal testis to be removed this request should not be granted. It should be pointed out to the owner that if they then tried to sell the animal as a gelding they could be found guilty of fraud or prosecuted under the Trade Description Act - both these are criminal affairs. There could also be the possibility of civil actions under the Sale of Goods Act and Misrepresentation Act. The veterinary surgeon who carried out the procedure can become involved in these actions and if he should be convicted in the courts of a criminal offence he runs the risk of appearing before the Disciplinary Committee of the Royal College of Veterinary Surgeons (Cox, 1973).

IV   ANAESTHESIA AND RESTRAINT

The horse should be starved of food for 24 hours prior to anaesthesia so as to reduce the bulk of abdominal contents, make a search for an abdominal tetis easier and reduce the danger of intestine becoming trapped in a laparotomy repair if a laparotomy should prove necessary.

If only one testicle is palpable the anaesthetic used should be capable of providing up to 45 min anaesthesia, although the length of time is only required if a laparotomy has to be performed. Chloral hydrate (14 gm/100 Kg bodyweight) supplemented, if necessary, by small doses of barbiturate, or induction with a short-acting barbiturate and maintenance with a volatile anaesthetic are both acceptable.

Whatever system is used, the horse should be placed in dorsal recumbency. This position is far superior to semi-dorsal recumbency and is easily realized and maintained under field conditions with a bale of hay or straw at each shoulder. The hind legs should be left completely free, in which case it is a good idea to cover the feet so as to prevent dirt falling off them onto the operation site.

V   PRE-OPERATIVE PREPARATION

Surgical instruments etc. for performance of a laparotomy should be available as well as equipment for castration. Examination of the inguinal region under anaesthesia may now show the presence of scars. The mere presence of a scar indicates no more than that somebody has made an incision - in one Leahurst case, a horse had five linear scars in the inguinal region and two abdominal testes. A <u>puckered</u> scar almost invariably indicates a scrotum was at one time incised, and by implication, that a testis has been removed.

Palpation of the inguinal region with the animal in dorsal recumbency may reveal a testis which was not previously palpable. Palpation is best carried out by pushing the fingers of the outstretched hand into the scrotal area, dorsally, laterally and slightly cranially towards the superficial inguinal ring. By pushing hard and deeply if necessary, structures inside the superficial ring can often be palpated by the finger tips. However, it must again be stressed not only that not all inguinal testes are palpable but also that other structures beside testes can be palpated in the inguinal region.

For these reasons, therefore, surgical exploration of the inguinal area is always imperative.

Accordingly, the scrotal area is clipped free of hair if this is necessary and if a definite testis is not palpable either on one or the other or on both sides, then on that side the ventral abdominal wall is clipped over a triangular area bounded by the midline medially, the line of the crutch caudo-laterally and a line joining the fold of the flank to a point 10 cm cranial to the umbilicus. This whole area is cleaned as for surgery.

Tetanus antiserum and intravenous crystalline penicillin are administered prophylactically prior to surgery.

VI   SURGICAL TECHNIQUE - INGUINAL EXPLORATION

The description which follows is based upon the author's experiences as outlined in Cox, Edwards and Neal (1975).

The surgeon stands or kneels behind the horse. If a testis can be brought to and held against the scrotal area it may be removed by either of the two methods described under castration. Otherwise an incision about 8 cm long is made through the skin in the area where the scrotum should be. (This approach has two advantages over an incision made over the superficial inguinal ring - it results in less haemorrhage from skin vessels and the inguinal fat does not have to be displaced or incised to locate the superficial inguinal ring). After incising the skin, the scalpel is laid aside immediately and dissection is continued towards the superficial ring with the fingers. This is necessary to avoid cutting the sometimes large branches of the external pudendal vein - these are usually larger and more obvious in older horses but they are present in young horses.

During dissection down to the inguinal canal, a search is made for structures other than fat and blood vessels. Four possible things may happen, (a), (b), (c) or (d).

(a)   A testis enveloped in a vaginal tunic may be found. It may then be brought to the exterior and removed.

(b)   A vaginal tunic without a testis may be found. This may vary in thickness from a fine pencil to a fat thumb and in length from very short and retained within the inguinal canal to long enough to be attached to the scrotum. It can be recognised by the whiteness of its fascial structure and by its having attached to it the cremaster muscle. In the horse in which a scrotal testis has been previously removed by an open castration technique, the distal end of the vaginal tunic is usually adherent to the scrotal scar. In cases in which no surgical interference has been made the vaginal tunic will only have tenuous fibrous connections with the scrotum. The vaginal tunic is in any case incised for about 5-10 cm and its contents inspected. Four possibilities exist, (i), (ii), (iii) and (iv).

(i) In the horse in which a testis has been previously removed, the deferent duct and remains of spermatic vessels will be found and will usually terminate in a fibrous nubbin in the region of the scrotal scar. No further action is necessary unless it is desired to remove a section of the spermatic sac for histological identification of remains of blood vessels and the deferent duct, in which case a 5 cm length may be readily obtained by use of the emasculator.

(ii) In the horse which is an incomplete abdominal cryptorchid, the deferent duct will again be found inside the vaginal tunic, but the duct will be seen to become epididymal tail distally and this in turn returns up the lumen of vaginal tunic as the epididymal body (Fig. 2:2). The latter two structures can be recognised because they consist of coiled tubes, the tail of the epididymis consisting of a large tube loosely coiled, and the body of the epididymis of a fine tube tightly coiled and recognisable only on close inspection. In such a case, traction may be applied to the body of the epididymis and it may be possible to deliver the testis through the vaginal ring and then sever its blood supply and attachments with the emasculator. Sometimes it is necessary to incise the vaginal tunic up to the level of the inguinal canal before the testis can be procured. It is rarely possible to deliver an abdominal testis in this way from a horse whose contralateral testis has been previously removed. If the testis cannot be delivered by traction, laparotomy is indicated.

(iii) Very rarely, a case of incomplete abdominal cryptorchidism is found in which the tail of the epididymis has been removed in the mistaken belief that it was a malformed testis. In such a case the vaginal tunic will contain deferent duct and body of epididymis passing distally to enter a nub of fibrous tissue.

Positive identification of the body of the epididymis allows this condition to be distinguished from the remains of the spermatic vessels. Traction on the body may deliver the testis, but if it does not laparotomy is indicated.

(iv) Occasionally a small vaginal tunic is found which contains only a single cord of fibrous tissue - this cord is the remains of the distal part of the true gubernaculum. In such a case, traction on this structure may deliver the tail of epididymis from the abdomen and by applying further traction to this body of the epididymis and eventually the testis can be delivered. Again, however, the vaginal ring may be too small to allow this to happen and a laparotomy is indicated to remove the testis.

(c) Sometimes no vaginal tunic is found even though adequate exposure of the superficial ring is obtained. The ring should be readily defined by its straight solid caudo-lateral edge and its crescent-shaped, softer but still tendinous, medio-cranial edge.

Two fingers should be inserted deep into the inguinal canal which passes laterally, dorsally and slightly cranially to ensure that no vaginal tunic is present within the canal. In such a case, the testis will be intra-abdominal and a laparotomy is indicated.

(d) Sometimes fibrosis from previous interference makes dissection impossible. In such a case either an immediate exploratory laparotomy is indicated or a blood test (p.43) for cryptorchidism should be performed to confirm or refute the presence of a testis.

If a scrotal testis is present on the opposite side, it should be removed prior to proceeding to a laparotomy. If no scrotal testis is present, on the opposite side, then exploration of the inguinal area should proceed along the lines outlined above.

VII SURGICAL APPROACH - LAPAROTOMY

A laparotomy is indicated (i) if the inguinal exploration has either established the presence of an abdominal testis or (ii) as an exploratory procedure because the inguinal findings are inconclusive.

Where unilateral abdominal cryptorchidism is present, the abdomen will be entered on the same side. In bilateral cases, the right-handed operator will find it more convenient to operate through the left-hand side. The surgeon stands or kneels alongside the horse.

A skin incision is made at about the level of the opening of the penile sheath and about 7-10 cm away from the mid-line. Local factors may dictate the exact site of the incision. It is wise to avoid large skin vessels if possible. It makes for an easier approach to the abdomen if much of the cod-fat can be avoided by making a slightly more cranial incision, but there is a risk that this will make exteriorization of the testis more difficult and should not be done in a case with bilateral abdominal retention. The line of the incision is made along the line of the fibres of the straight abdominal muscle. Therefore, when the incision is made level with or cranial to the sheath opening it runs parallel to the mid-line. Made further caudally, the incision must be angled slightly medially as it passes caudally because the fibres are converging on the linea alba in that region. The incision is made long enough to accommodate the surgeon's whole hand.

The fat of the underlying superficial abdominal fascia ('Cod Fat') is often thicker caudally than cranially, and is often traversed by large veins which, because of the dense nature of the fat, are not always seen before they are incised. Once through this fat layer, however, haemorrhage is rarely a problem. The abdominal tunic and closely adherent tendons of the external and internal oblique muscles are incised with a scalpel and the underlying straight abdominal muscle split along the length of the incision by blunt dissection. Occasionally minor haemorrhage occurs from the muscle but the result is usually a blood-free view of the transverse tendon. Fig. 2:3 shows the relationships of all these structures in this region. The whole hand is introduced into the abdomen and a search made for the testis. Most of the testes will be found readily close to the vaginal ring. For locating a testis not close to the vaginal ring, it is useful to identify the deferent duct where it lies dorsal to the bladder, and to follow this to the tail of the epididymis, thence to body and head of epididymis and finally to the testis (Fig. 2:1). This technique fails when excessive fat makes tracing the duct difficult and is useless in cases of incomplete abdominal cryptorchidism. An alternative procedure is to fan the hand about in the abdomen until the testis falls into it. This is surprisingly successful and avoids extensive handling of the intestine - it does not, however, always work!

Although most testes are found close to the deep inguinal ring, they may be dorsal to the rectum, immediately dorsal to the incision, near the kidney or near to or actually within the pelvic inlet alongside neck of bladder. One of the most important single factors in accurate localisation of the testis is recognition of the flabby nature of the majority of abdominal testes - the person unaware of this fact may fail to realise that he has in fact grasped the testis. Very small testes or large cystic testes may present above average difficulty in detection. Once grasped the testis is brought out through the incision and its blood vessels severed by an emasculator, left in place for at least 2 minutes - the application of a separate ligature is not necessary. In many cases it is impossible to exteriorise completely the epididymis through the laparotomy incision and in most cases the emasculator should merely be applied as high up as possible. In the case of a bilateral abdominal cryptorchid, it is usually possible to remove both testes through the one incision, although difficulties of exteriorisation may mean that a ligature and scissors or an ecraseur have to be used for haemostasis and removal, or (rarely) a laparotomy incision on the other side.

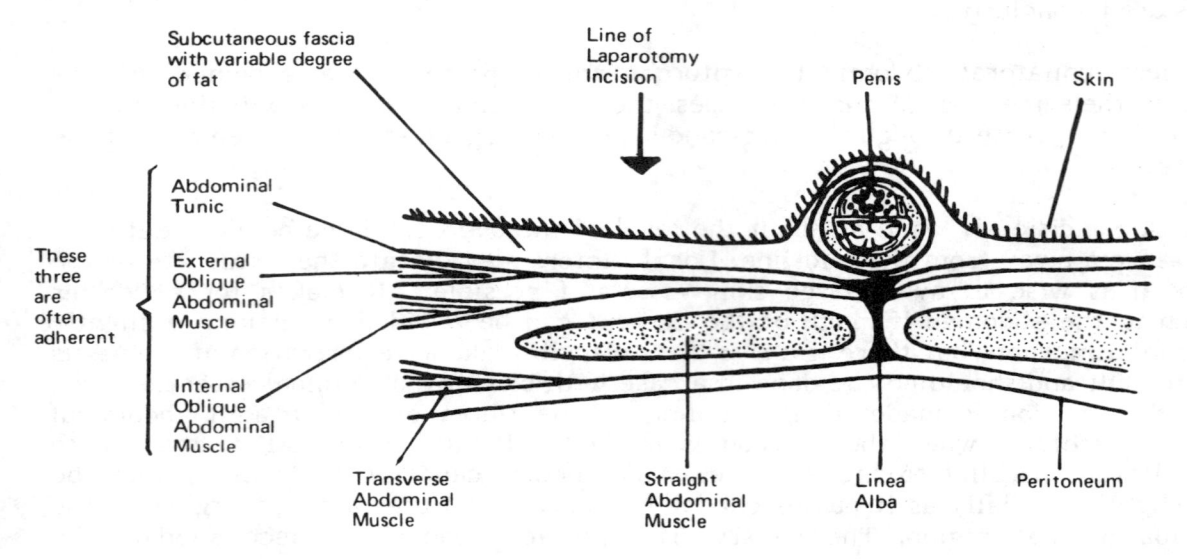

Fig. 2:3 Layers of abdominal wall of the horse in transverse section in the region of a supra-pubic paramedian laparotomy

The incision is repaired in several layers. Firstly, one single interrupted suture is placed through the central points of the breaks in the straight abdominal muscle and the transverse tendon - this is placed down through the straight muscle on the distal side, down through the cranial side of the transverse tendon, up through the caudal side of the transverse tendon and up through the proximal side of the straight abdominal muscle. Thus, the cruciate approach to the abdomen in this area is sealed at the area where the two lines cross. Care must always be taken that the intestine is not trapped in this suture. A further one or two interrupted sutures are placed on either side of this suture through the straight abdominal muscle. The incision in the abdominal tunic and closely adherent tendons of the two oblique muscles is repaired by interrrupted or by horizontal mattress sutures, the loose connective tissue and fat apposed as necessary the skin incision closed with horizontal mattress sutures. Sutures of synthetic absorbable material are used throughout.

VIII POST-OPERATIVE COMPLICATIONS

The post-operative complications which may follow the inguinal exploration are similar to those which may follow a routine castration (see pp. 16ff).

However, because dissection has been carried out deep into the inguinal region there is an increased degree of trauma and this sometimes results in post-operative oedema. The most serious complication is if one of the pudendal veins is severed - in order to avoid this possibility the scalpel is discarded after the skin has been incised and all dissection is made bluntly.

Following the supra-pubic paramedian laparotomy described above, certain complications may arise:-

(a) Haemorrhage. Experience with over 250 operations has shown that haemorrhage from the spermatic vessels is not a problem - the emasculator alone without a ligature produces effective haemostasis. More problems are caused by haemorrhage from vessels in the subcutaneous fascia and careful attention to them is necessary.

(b) Sepsis. Post-operative sepsis has not been a problem with synthetic absorbable sutures, although it was with catgut.

(c) Oedema. Synthetic absorbable sutures produce much less post-operative oedema than catgut, although the degree of trauma involved in entering the abdomen and removing a testis also contributes to the development of oedema. The oedema can produce stiffness and unwillingness to walk but both the stiffness wears off and the oedema dissipates if the animal is made to exercise.

(d) Colic. Occasionally mild colic develops a few hours after operating but this is readily cured by the administration of pethidine. The cause of the colic is obscure but it may be due to anaesthesia rather than the handling of intestine. One horse at Leahurst developed severe colic several days after laparotomy and this was shown to be due to intestine trapped in the suture line. This case underlines the need (i) to starve the animal adequately prior to surgery and (ii) to ensure that intestine is not trapped in the first suture.

(e) Dullness. Occasionally dejection persists into the post-operative days, possibly due to mild peritonitis. A four day course of antibiotic administration usually results in a rapid improvement.

IX DISCUSSION OF APPROACH TO ABDOMINAL CRYPTORCHID

The technique of supra-pubic paramedian laparotomy was first described as long ago as 1838 but fell into disuse until it was revived by Wright in 1955. Since that date it has been used at Leahurst in over 250 abdominal cryptorchids with only one serious surgical complication (the horse with gut trapped in the suture line which was mentioned above), and one death (in a horse recovering from anaesthesia in which the cause of death was not determined). The technique has been adopted as standard by several other U.K. schools and by veterinary schools in Holland, Germany and the U.S.A. Nevertheless, other approaches are used and have their advocates.

An alternative laparotomy site is the sub-lumbar region, first described in 1860, but this has little to commend it over the supra-pubic site except that the operation may be carried out with the horse standing (!). Its disadvantages compared with the supra-pubic approach are (i) seromas are more of a problem, (ii) the bilateral case is less easily dealt with and (iii) the horse must be rolled from dorsal recumbency into lateral recumbency during the course of the operation. An account of the technique will be found in Arthur (1961).

The most popular alternative to the supra-pubic route is an approach through or close to the inguinal rings and a variety of different techniques are described (Adams, 1964; Bishop and others 1964; O'Connor, 1938). Many, however, carry with them the risk of post-operative intestinal prolapse and, although in the hands of the experienced operator that risk may be small, in inexperienced hands that risk can be considerable. Moreover, location of the testis which does not lie close to the vaginal ring is not always easy, removal of large testes is difficult and the animal suffers more post-operative discomfort than if the abdomen is entered elsewhere. In addition, the rate of technical failure especially in inexperienced hands is high (Stickle and Fessler, 1979).

In conclusion, the approach to the abdominal cavity through a supra-pubic paramedian laparotomy is easy, the testis itself is generally readily located and removed whatever its size, and controlled repair of the incision is possible. Post-operative complications are rare and generally not serious and the dangers of post-operative prolapse of bowel, even in inexperienced hands, almost entirely eliminated.

C. REFERENCES

Adams, O.R. (1964) An improved method of diagnosis and castration of cryptorchid horses. Journal of the American Veterinary Medical Association 145 : 439-446.

Arthur, G.H. (1961) The surgery of the equine cryptorchid. Veterinary Record 73 : 385-389.

Arthur, G.H. (1975) Veterinary Reproduction and Obstetrics. Fourth Edition. Bailliere and Tindall, London.

Bergin, W.C., Gier, H.T., Marion, G.B. and Coffman, J.R. (1970) A developmental concept of equine cryptorchidism. Biology of Reproduction 3 : 82-92.

Bishop, M.W.H., David, J.S.E. and Messervey, A. (1964) Some observations on cryptorchidism in the horse. Veterinary Record 76 : 1041-1048.

Cox, J.E. (1973) The castration of horses : or castration of half a horse? Veterinary Record 93 : 425-426.

Cox, J.E. (1975) Experiences with a diagnostic test for equine cryptorchidism. Equine Veterinary Journal 7 : 179-183.

Cox, J.E. (1986) The behaviour of the false rig - its causes and consequences. Veterinary Record 118 : 353-356.

Cox, J.E., Edwards, G.B. and Neal, P.A. (1975) Supra-pubic paramedian laparotomy for abdominal cryptorchidism in the horse. Veterinary Record 97 : 428-432.

Cox, J.E., Edwards, G.B. and Neal, P.A. (1979) An analysis of 500 cases of equine cryptorchidism. Equine Veterinary Journal, 11 : 113-116.

Cox, J.E., Redhead, P.H. and Dawson F.E. (1986) A comparison of the measurement of plasma testosterone and plasma oestrogens for the diagnosis of cryptorchidism in the horse. Equine Veterinary Journal 18 : 179-182.

Foster, A.E.D. (1952) Polyorchidism. Veterinary Record 64 : 158.

Hayes, H.M. (1986) Epidemiological features of 5009 cases of equine cryptorchidism. Equine Veterinary Journal 18 : 467-471.

Hobday, F.T.G. (1914) Castration (including Cryptorchids and Caponing) and Ovariotomy. Edinburgh, W. Johnston.

Leipold, H.W., De Bowes, R.M., Bennett, S., Cox, J.H. and Clen, M.F. (1986) Cryptorchidism in the horse : genetic implications. Proceedings of the Annual Convention of the American Association of Equine Practitioners 31 : 579-590.

Lush, J.L., Jones, J.M. and Dameron, W.H. (1930) The inheritance of cryptorchidism in goats. Bulletin of the Texas Agricultural Experimental Station 407.

Mikami, H. and Fredeen, H.T. (1979) A genetic study of cryptorchidism and scrotal hernia in pigs. Canadian Journal of Genetics and Cytology 21 : 9-19.

O'Connor, J.J. (1938) Dollar's Veterinary Surgery. General, Operative and Regional. Third Edition. Bailliere, Tindall and Cassell, London.

O'Connor, J.P. (1971) Rectal examination of the Cryptorchid Horse. Irish Veterinary Journal 25 : 129-131.

Stanic, M.N. (1960) Castration of Cryptorchids. Modern Veterinary Practice 41 : 30-33.

Stickle, R.L. and Fessler, J.F. (1978) Retrospective study of 350 cases of equine cryptorchidism. Journal of the American Veterinary Medical Association 172 : 343-346.

Wright, J.G. (1963) The surgery of the inguinal canal in animals. Veterinary Record 75 : 1352-1363.

Chapter Three
HERNIAS AND RUPTURES IN THE INGUINAL REGION

A. ANATOMICAL CONSIDERATIONS

I DEFINITIONS

A hernia or rupture is a protrusion of an organ or part of an organ or other structure through the wall of the cavity normally containing it. A hernia or rupture consists of three parts, viz. a ring, a sac and contents:

(i) The ring is that opening in the cavity wall through which the hernia or rupture occurs.

(ii) The sac encloses the contents of the hernia or rupture. The sac may be divided for convenience into a neck (that part closest to the ring), a body (that part between neck and fundus) and a fundus (that part furthest from the ring).

(iii) The contents of a hernia or rupture involving the abdominal wall are usually intestine and mesenteric support or omentum.

The words hernia and rupture are often considered to be synonyms and some authorities even state that hernia is Latin for rupture. However, both 'hernia' and 'ruptura' occur in Latin and a clinically useful distinction can be drawn between the two words. Rupture can be used to describe a situation where discontinuity of tissue has occurred to form an unnatural ring. Hernia can be used where the defect forming the ring is natural, even if only in the foetus.

Usually, therefore, a hernia has a sac whose innermost layer is peritoneum whereas in a rupture the peritoneal continuity is disrupted and the sac is formed solely by local fascia. It would thus be proper to talk of diaphragmatic rupture in a cat following a road accident but of an umbilical hernia in a newborn calf. Moreover, a hernia is usually (but not always) congenital whereas a rupture is almost invariably acquired.

Hernias and ruptures may be classified as to whether or not the contents can be returned to the cavity by simple manipulation. If the contents can be so returned, the hernia or rupture is said to be reducible. Irreducible ones may be divided into three types. The first is one incarcerated by distension, in which the contents (usually intestines) are too distended to be returned by simple manipulation but no other pathology is present. The second type is one incarcerated by adhesion, in which the contents are adherent to the lining of the sac. In some cases both distension and adhesion may be present. The third type is strangulated and in these the blood supply to the contents is compromised, either by the ring or by the neck of the sac or both.

Distinctions are sometimes made between inguinal hernias and ruptures, in which the contents reach only the inguinal canal, and scrotal hernias and ruptures, in which the contents reach the scrotum. However, the superficial inguinal ring and the scrotal neck are always separated by a considerable distance, so how is one to define a hernia whose contents have passed beyond the inguinal canal but not reached the scrotum? Here, therefore, all hernias and ruptures are defined by the area through which the protrusion occurs, viz. as inguinal hernias or ruptures.

II THE ANATOMY OF THE INGUINAL REGION

The most important concept to grasp in considering the anatomy of the inguinal region is the structure of the inguinal canal. This canal can be considered as a short tube passing through the abdominal wall. The limits of this tube are the inguinal rings, the deep (internal) inguinal ring and the superficial (external) inguinal ring.

The various layers of the body wall which define the inguinal canal have curved surfaces and they are, therefore, notoriously difficult to represent adequately in essentially two dimensional drawings. The best way to understand this area is to dissect it carefully on cadavers.

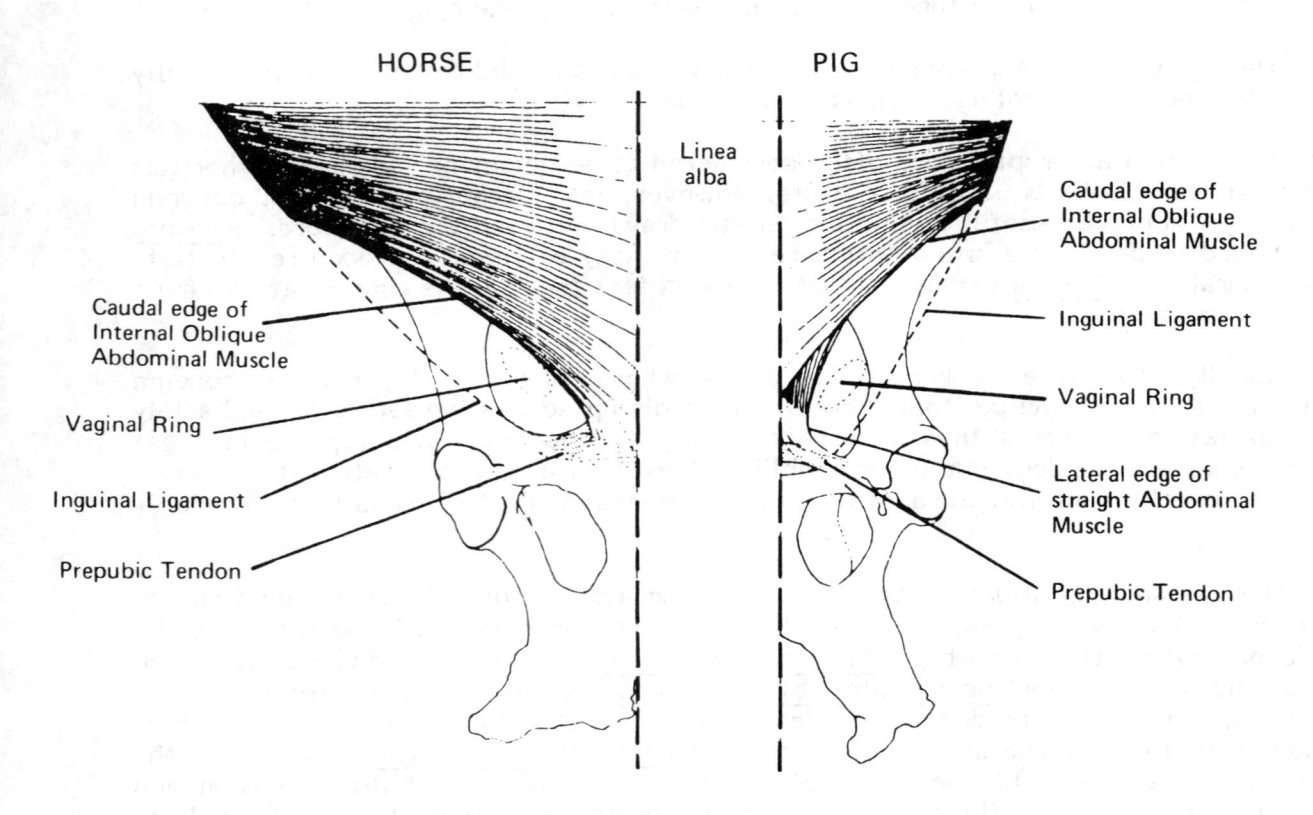

Fig. 3:1 Ventral view of inguinal region in horse (left) and pig (right) — internal oblique abdominal muscle.

(a) The Deep Inguinal Ring

In the horse this structure is bounded ventro-cranially by the caudal edge of internal oblique abdominal muscle and dorso-caudally by the inguinal ligament which is the caudal edge of the pelvic tendon of the external oblique abdominal muscle (Fig. 3:1L).

In the pig, the internal oblique abdominal muscle does not insert onto the prepubic tendon or the most caudal section of the linea alba and in this species, therefore, the ventral edge of the deep inguinal ring is formed by the lateral edge of the straight abdominal muscle and the prepubic tendon (Fig. 3:1R).

The ruminant species occupy an intermediate position as the insertion of their internal oblique muscles extends more caudally than in the pig but is not as extensive as in the horse.

The deep inguinal ring is mostly occupied by a layer of fascia, a caudal continuation of the fascia of the internal oblique muscle. Deep to this lies the transverse fascia and then peritoneum.

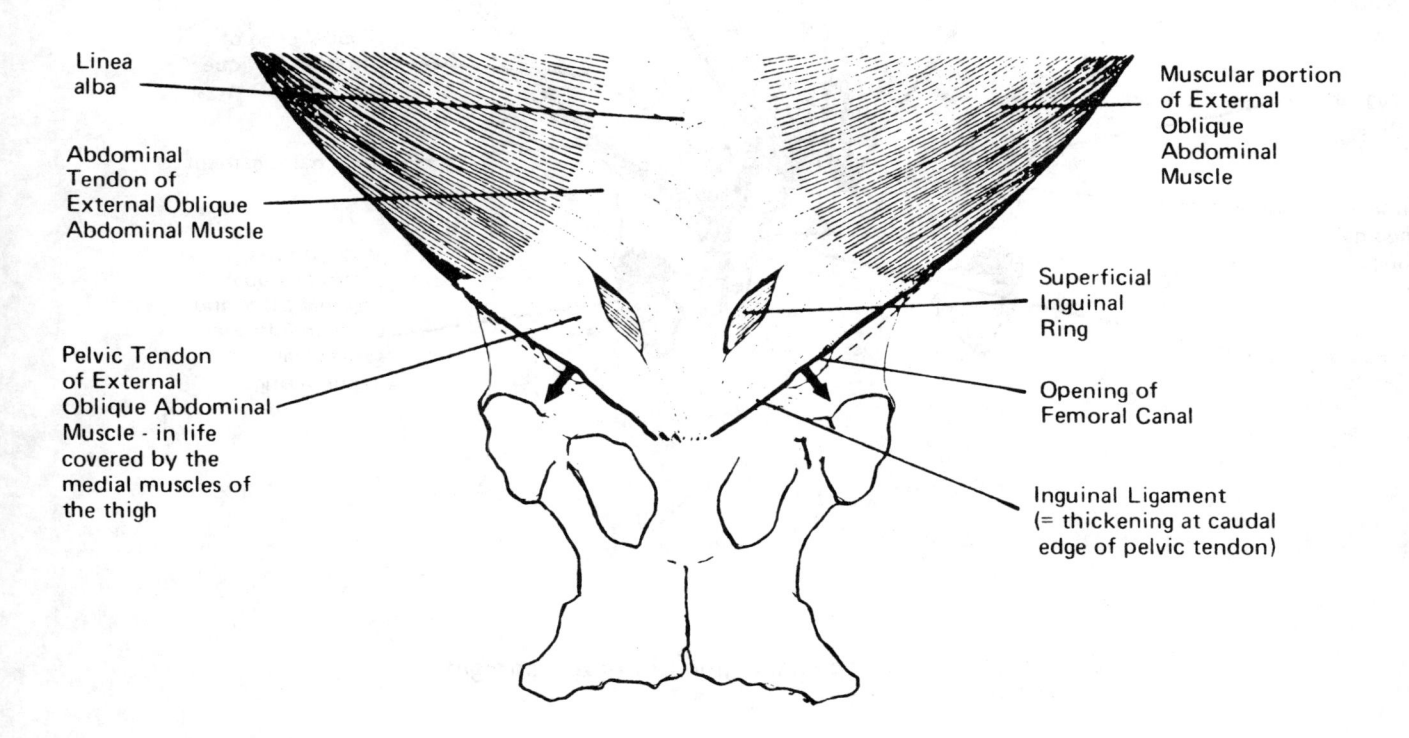

Fig. 3:2 Ventral view of inguinal area of horse — external oblique abdominal muscle.

(b) **The Superficial Inguinal Ring**

This is in all species a slit-like opening in the external oblique abdominal tendon. The slit runs along the line of the fibres from cranio-dorsally to caudo-ventrally and it separates the tendon into a cranial abdominal portion and a caudal pelvic portion (Fig. 3:2).

In the horse the ring is readily palpable in the groin as an arc, larger in the anaesthetised horse on its back than in the conscious horse. The curve lies cranially and is the caudal edge of the abdominal tendon of the external oblique abdominal muscle. The straight part is formed by the cranial edge of the pelvic tendon supported by the medial muscles of the thigh caudo-laterally and the tough inguinal ligament dorsally, and thus feels very hard and solid.

In the pig, the superficial inguinal ring is palpable as a slit, just cranial to the junction of body wall and thigh.

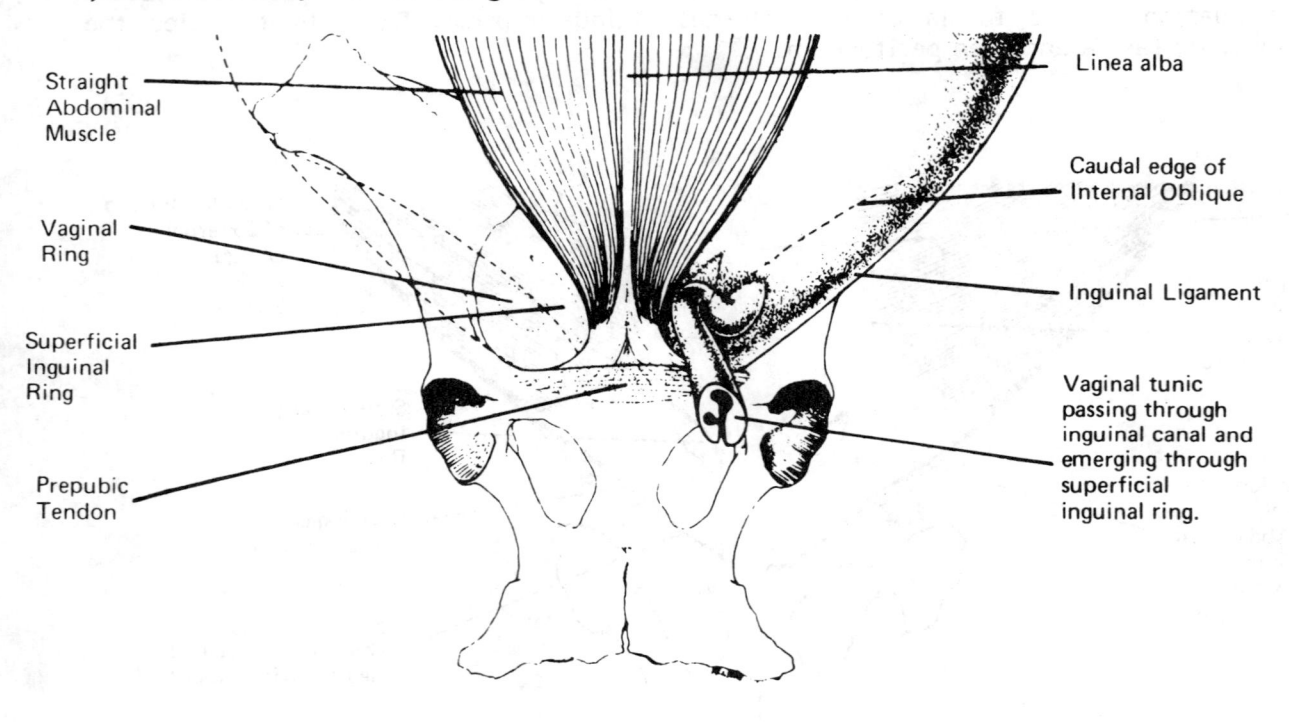

Fig. 3:3 Ventral view of inguinal region of horse — straight abdominal muscle.

(c) **Inguinal Canal**

In the horse, the two inguinal rings do not overlie each other, their ventro-caudal angles lying about 4 cm apart, their cranio-dorsal angles some 17 cm apart. The inguinal canal (Figs. 3:3 and 3:4) is thus much shorter ventro-caudally than cranio-ventrally. Its lateral wall is the pelvic tendon of the external oblique muscle and its medial wall the internal oblique muscle. In this species, therefore, as in man, the tension of the abdominal muscles tends to keep the canal almost closed, the internal oblique muscle acting like a flap over the superficial ring.

In the pig, the deep and superficial inguinal rings actually overlie each other and in this species, therefore, the canal is virtually non-existant (Fig. 3:5). In the ruminants the canal is shorter than in the horse because the internal oblique muscle does not reach so far caudally.

In both sexes the canal is occupied by the external pudendal artery and the inguinal lymph vessels which drain the preputial and penile region of the male and, in part, the udder of the female. In the ruminant and pig the external pudendal vein passes up the inguinal canal. (In the horse, however, the main vein passes through a foramen in the cranial part of the tendon of origin of the gracilis muscle and thus passes into the abdominal cavity caudal to the inguinal canal whilst only a small vein accompanies the external pudendal artery through the canal). These structures enter the canal caudally in the deep inguinal ring and emerge at the medial angle of the superficial ring.

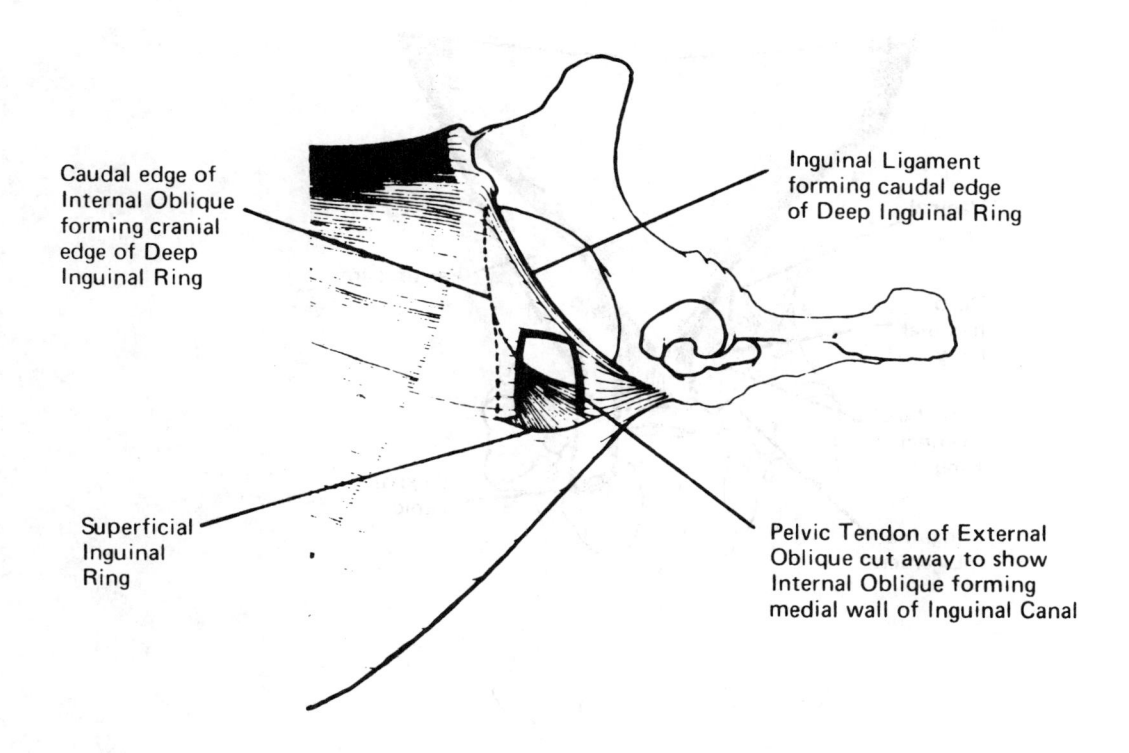

Fig. 3:4 Lateral view of inguinal region of the horse to show the inguinal canal.

In the entire male the inguinal canal is distended by the neck of the spermatic sac (see p.8) as shown in Figs. 3:3. It is this distension which predisposes the male to congenital and acquired inguinal hernia. After castration the spermatic sac undergoes some atrophy, and so in the castrated male the canal is much less distended and herniation is, therefore, rare. In the female of the species with which this book is concerned, no vaginal process invades the inguinal canal to produce or allow any dilation and so congenital or acquired inguinal hernia is also rare. In the female, however, the prepubic tendon (Fig. 3:1) may degenerate and a rupture ensue.

(d) The Vaginal Ring (Figs. 3:1, 3:3 and 3:5)

The vaginal ring is the opening of the lumen of the vaginal tunic into the abdominal cavity. It is, therefore, present only in the male and it occupies the most dorsal area only of the deep inguinal ring. It is not actually a circular shaped opening but is anatomically more like a C (see p.3) and clinically more like a slit, than a circular opening.

It may be palpated in the tranquillized horse as follows (O'Connor, 1971): the left hand is put into the rectum to palpate the right ring, and vice versa: The lateral aspect of the wrist is rested on the pelvic brim with the finger tips pressed against the abdominal wall laterally; the middle finger is flexed and then, as it is extended, it should enter the vaginal ring. A flap of peritoneum will cover the entrance to the ring rendering it impalpable if the fingers are drawn backwards over it.

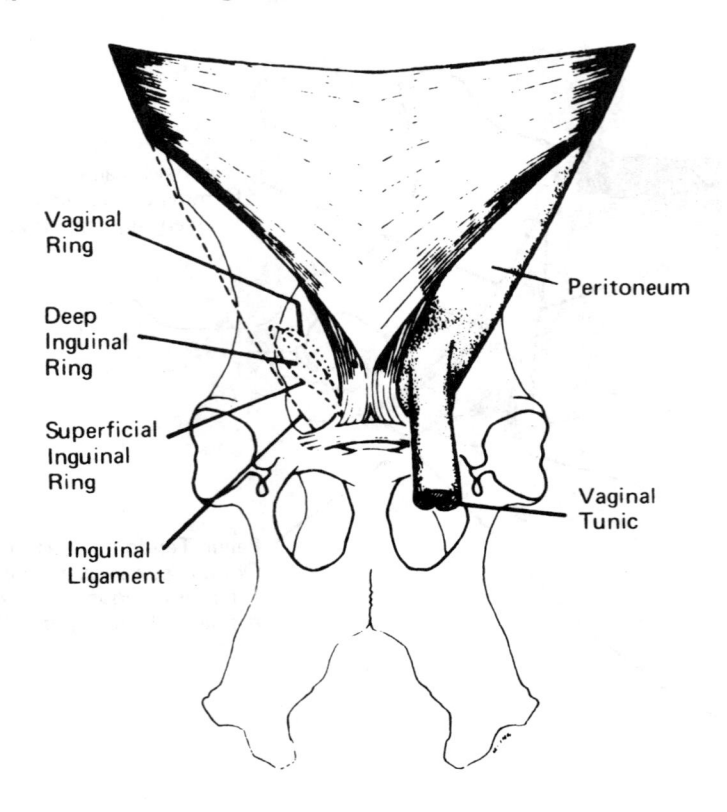

Fig. 3:5 Ventral view illustrating the inguinal canal of the pig.

(e) Difficulties

Confusion arises in discussing this area anatomically or clinically because the vaginal ring and the deep inguinal ring are sometimes considered to be synonyms, whereas the former is only a small area of the latter and is present only in the male. Moreover, the term 'inguinal canal' may be used as though it is the same as the 'spermatic sac', whereas the canal is occupied by more than the spermatic sac. In addition, the area between the superficial inguinal ring and the neck of the scrotum is sometimes spoken of as being part of the inguinal canal - this is anatomical nonsense as the external limit of the canal is the superficial ring. This area should be called the inguinal region.

B. CLINICAL CONSIDERATIONS

In the domestic animals, the commonest form of inguinal hernia forms through the vaginal ring, which is thus the hernial ring. The sac of the hernia is the vaginal tunic, the contents of the hernia occupying its lumen (Fig. 3:6). Occasionally, a pouch of peritoneum may develop alongside the vaginal tunic but independant of it, but this form of inguinal hernia, although common in man, is extremely rare in the domestic animal.

Ruptures can occur in association with the spermatic sac. The exact location of the rupture can be difficult to determine exactly at surgery, but the most likely possibility is in the neck of the sac as it passes through the inguinal canal. The prolapsed intestines then come to lie outside the spermatic sac in the relatively loose spermatic fascia. They may only just pass through the deep ring or may reach as far as the scrotum.

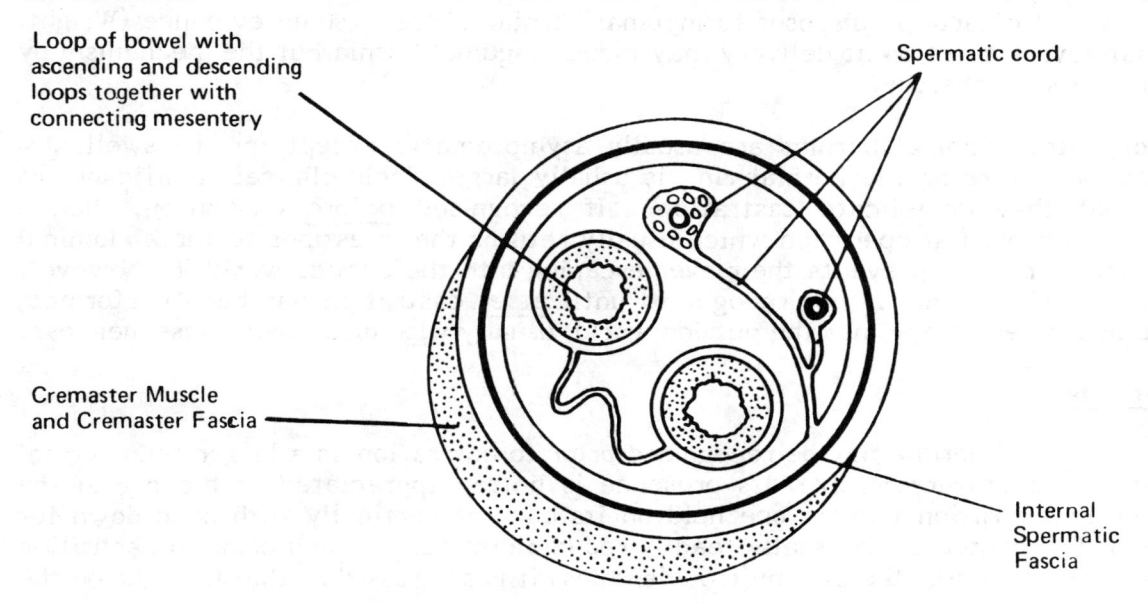

Fig. 3:6 Diagrammatic Cross Section of Spermatic Sac with loop of bowel in Lumen of Vaginal Process — compare Fig. 1:3.

I   CONGENITAL INGUINAL HERNIA

(a)   General Considerations

There is good evidence that congenital inguinal hernia is inherited. The condition is extremely rare in horses in the U.K., almost certainly because the Horse Breeding Act forbad the licensing of a stallion with the condition. Warwick (1926) concluded from breeding experiments with pigs that the condition was caused by a double recessive genotype. It has also been shown that there is a correlation between inguinal hernia and cryptorchidism in the pig (Wensing and Colenbrander, 1973; Mikami and Fredeen, 1979) and the genetic predisposition to both has been established by the latter authors. In the author's opinion, therefore, animals with congenital hernia should always be destined for castration and should not be used as breeding animals.

The virtual absence of the inguinal canal which was noted above (p.57) is probably the factor which makes congenital hernia occur most commonly in the pig amongst the domestic animals, although Warwick (1926) has argued that it is the size of the vaginal ring in the pig which predisposes this species to herniation. Sporadic cases of inguinal hernia probably occur in all pig units.

Wensing and Colenbrander have also shown that the development of inguinal hernia in the pig is closely related to the extent to which the gubernaculum dilates the inguinal canal during the process of testiscular descent. It seems likely, therefore, that one of the reasons congenital inguinal hernia occurs next most commonly in the horse is that in this species the testis has descended only just before birth (p.4) and thus, at birth, the inguinal canal has only recently been dilated by the gubernaculum. In some congenital cases of inguinal hernia in the horse the internal oblique muscle does not completely underlie the superficial inguinal ring (unpublished observations). Whether this anatomical abnormality is caused by abnormal gubernacular development or not is unknown, but it clearly predisposes to inguinal hernia. There is some evidence (Wright, 1963) that severe traction at delivery may induce inguinal hernia but the mechanism by which it does so is obscure.

Congenital inguinal hernias are usually asymptomatic except for the swelling - strangulation is rare as the hernial ring is usually large. Their clinical significance is chiefly that they complicate castration. If recognised before castration, then a technique is adopted at operation which usually returns the intestines to the abdominal cavity and, therefore, prevents their ever escaping into the outside world. If, however, the presence of a hernia is not recognised until after castration has been performed, then the intestines escape into the outside world, usually with disastrous consequences.

(b)   The Pig

In the pig inguinal hernia may be recognised prior to castration as a larger than normal scrotal swelling of varying size. Its presence is better appreciated in the pig in the normal upright position than in one held on its back or vertically with head down for castration. The contents are usually readily reducible at surgery and prior to castration the hernia rarely causes the pig any trouble unless it is so gross that the skin rubs on the floor.

(c)   The Horse

In the foal, the presenting sign is unilateral or bilateral scrotal swelling. The swelling may not be very large and may not inconvenience the foal at all (but see inguinal ruptures, page 64). Provided, therefore, the foal remains fit and continues to grow well, the hernia may be left and, in the great majority of cases, spontaneous resolution will have occurred by 12 months of age. As the foal and his intestines grow larger, the inguinal canal becomes smaller and the mesentery, apparently, shorter and the intestines are gently, but inexorably, returned to the abdominal cavity by peristaltic activity. Sometimes, however, the swelling may be large enough to inconvenience the foal in walking or galloping, or so large as to impede lymphatic drainage from the preputial area, which thus becomes oedematous. In such a case, or if there is deterioration in the foal's condition or he fails to grow as quickly as he should, the hernia should be repaired immediately. Although it seems possible that the condition resolves completely and the inguinal rings become of normal size, in some cases the inguinal rings may remain slightly enlarged and abdominal contents herniate more readily when the animal is castrated. Care should always be taken, therefore, to perform a 'closed' castration (see p.13) on a horse with a history of inguinal hernia as a foal. (see also p.64 on inguinal rupture in foals)

Inguinal hernia in the <u>older</u> horse may only be noticed when the horse is presented for castration. Although it is uncommon in most countries and in most breeds (except possibly the standardbred), every horse presented for castration should be examined for herniation prior to surgery. If the condition is unilateral, one scrotum is larger than the other, a feature especially noticeable if the horse is viewed from behind with the tail raised. If a hernia is present on both sides, both scrotums will be seen to be larger than normal for the age of the animal. Careful palpation of the scrotum will reveal that something other than testes and epididymis is present in the vaginal sac and rectal examination will enable intestines to be felt as they pass through the vaginal ring. (Rectal palpation of the vaginal ring prior to routine castration may indicate the possibility of herniation, but will not infalllibly pick out horses which will herniate after castration, as in many such cases the vaginal ring is of normal dimensions). The scrotal swelling of a hernia must be differentiated from (i) hydrocoel (only fluid is present in the vaginal sac - see p.33), (ii) haematocoel (rare), (iii) cystic ends to cords (a fluid-filled sac at the end of the spermatic cord; usually a complication of castration - see p.20), (iv) scrotal oedema (which pits on pressure) and (v) scrotal enlargement due to torsion of the spermatic cord (see p.34).

(d) <u>The Ruminant</u>

Congenital inguinal hernia is extremely uncommon in any of the ruminant species. A soft fluctuating swelling, most marked in the scrotal neck, will be palpable but must be differentiated from testicular swelling, epididymal swelling, excess fluid and excess cod fat (Arthur, 1963 - see Ashdown, 1963).

II   ACQUIRED INGUINAL HERNIA

(a) <u>General Considerations</u>

Cases of acquired inguinal hernia almost always appear to be associated with some traumatic incident, but the mechanism by which the traumatic incident results in herniation is obscure. Two factors may be involved:

(i) <u>Increase in Abdominal Pressure</u>. Lifting of heavy loads is believed to be an important factor in the development of inguinal hernia in the adult human male, as it leads to a marked increase in abdominal pressure. The inguinal area may be considered as a weak point, equivalent to a safety valve, so herniation occurs through it if the abdominal pressure becomes too great.

(ii) <u>Integrity of the Inguinal Canal</u>. Whether or not the inguinal canal dilates sufficiently to allow intestines into the lumen of the vaginal tunic depends at least in part upon the activity of the internal oblique muscle which, in the horse and to a lesser extent in the ruminant forms the medial wall of the canal. If the flap-valve action of this muscle fails then herniation will occur more readily. There is evidence from man (Radojevic, 1962 - see Ashdown, 1963) that there is considerable individual variation in the anatomy of the inguinal area. The same is probably true within one species of domestic animal, perhaps with certain breeds (e.g. Scottish Blackface Rams, and Standardbred Horses -see below) having an anatomy which is particularly vulnerable.

There are differences in the clinical signs shown by the different species so each will be considered in turn.

(b)   The Ram

Acquired inguinal hernia is uncommon in sheep except in Scottish Blackface Rams (Orr, 1958). It is usually a sequel to an episode of fighting which, in this breed, involves head-on-collisions. There is presumably an enormous sudden increase in abdominal pressure at the moment of impact which is the immediate cause of the herniation. However, it has not been determined whether there is a congenital weakness of the structures of the inguinal canal in this breed (Arthur, 1956) or whether herniation is merely a reflection of the venom of this breed's fighting. The presence of herniated contents in the vaginal tunic occludes the blood supply and venous and lymphatic drainage from the testis. The testis on the affected side (more commonly the right -the condition is rarely bilateral), therefore, swells initially and then undergoes shrinkage due to atrophy. The swelling in the body and fundus of the spermatic sac not only produce an abnormally swollen scrotum but also produce a stiffness in gait - indeed, it is often this stiffness which directs the stockman's attention to the ram. Palpation confirms that the scrotal changes are due to herniation and not to orchitis or direct trauma to the scrotum.

The contents are intestine and omentum, most usually the latter, although the bladder has also been found. Strangulation is rare, perhaps because the deep inguinal ring in this species is large enough to avoid the severe constriction which occurs in the horse (see below).

(c)   The Bull

Acquired inguinal hernia is extremely rare in the bull although cases associated with trauma or following service do occur.

(d)   The Stallion

Acquired inguinal hernia is extremely rare in the stallion. It was seen not infrequently in the stallions of the Indian cavalry at the turn of the century (Meredith, 1891). There may be a history of becoming cast, of falling awkwardly ('coming a purler') or of trying to jump a fence. Any one of these situations might produce either sudden increases in abdominal pressure or inco-ordinate movements which affect the flap-valve action of the internal oblique muscle or both.

In a case of inguinal hernia in the stallion, the intestine usually becomes trapped in that part of the vaginal tunic which lies within the confines of the inguinal canal, so symptoms of bowel obstruction develop. The intestine becomes unable to pass on gas or ingesta by peristalsis and so becomes distended. The distension embarrasses the blood supply to the herniated loop, i.e. strangulation occurs. Eventually ischaemia develops and necrosis ensues. Perforation may result but the horse usually dies before this occurs. These changes occur quickly, generally within a matter of hours, probably becaue the inguinal canal is hardly distensible and the intestines are effectively and completely trapped. The clinical signs of horses with intestines strangulated in the inguinal canal have been well described by Meredith (1891). The temperature may be slightly elevated, normal or sub-normal, the pulse rate and respiratory rate accelerated. The animal shows considerable uneasiness, getting up and lying down frequently, perhaps standing with arched spine and characteristically looking round at the flank of the affected side (the condition is usually unilateral). Sometimes the horse may stretch out his head and neck with the upper lip turned up. The flank, loins, thigh and scrotum of the affected side may be covered with an icy sweat and the scrotum appears swollen and may sometimes be cold to the touch. Palpation of the scrotum may confirm that the swelling is due to intestine.

Acquired inguinal hernia should <u>always</u> be considered as a cause of colic in stallions but the condition needs to be differentiated from torsion of the spermatic cord - see p.34 - and rectal examination will confirm the presence of herniation in the inguinal area. Immediate surgery is indicated to relieve the obstruction. Prognosis improves with prompt recognition and surgery (see also Schneider, Milne and Kohn, 1982).

(e)   Related to Castration in the Horse

Acquired inguinal hernia may also develop in the horse during or shortly after castration. Whilst some horses in which intestine prolapses after castration may have had congenital inguinal hernia which was missed in the pre-operative examination, in many instances there is no clinical evidence of a hernia prior to operation and the vaginal ring and inguinal canal are of normal size.

(i)   Herniation during Castration

Day (1966) records that, when colts were castrated in the standing position without any local anaesthesia but with the tail forcibly elevated, abdominal straining resulted and occasionally prolapse of intestine or omentum occurred whilst the castration was being performed. Intestine or omentum may also be found in the lumen of the vaginal tunic when castration is carried out under general anaesthesia. Unpublished observations suggest that this situation arises more frequently when the neuroleptanalgesic 'Immobilon' is used. It seems likely that an increase in abdominal pressure occurs with 'Immobilon' and forces a loop of bowel, or a piece of omentum, through the vaginal ring.

(ii)   After Castration

Prolapse of bowel or omentum may occur in horses recovering from a general anaesthetic administered for castration. Occasionally, however, herniation may not occur for several days. In these cases it may be that abdominal straining during attempts to rise has produced a marked increase in abdominal pressure, so resulting in the herniation, or that inco-ordinate movements during attempts to rise have affected the integrity of the flap-valve action of the internal oblique muscle medial to the superficial inguinal ring and so allowed herniation to occur. Hutchins and Rawlinson (1972) have presented some evidence suggesting that this is more common in the older Standardbred ('Trotting') horse, but were unable to reach any certain conclusions as to why this should be so.

In any case of acquired inguinal hernia after castration the pathology is that of the acquired inguinal hernia in stallion described above under (b) compounded by exposure of intestines with drying, damage from being trodden on, etc. Immediate surgery or destruction on humane grounds is indicated, according to the extent of damage to the intestines, the state of the horse as a whole, and the facilities available for correction of the condition.

If, however, the contents of the hernia are only omentum then the situation is not serious as the omentum effectively plugs the vaginal ring - it is only necessary to cut off the prolapsed piece of omentum just inside the scrotal wound.

Occasionally prolapse of vaginal tunic alone through the scrotal incision occurs, giving the impression at first glance that a small loop of intestine is prolapsing. The long vaginal tunic of the cob and shire seems particularly likely to do this. The 'prolapse' usually disappears when the effects of anaesthetic and tranquillisers wear off.

III   INGUINAL RUPTURE

In this condition, prolapse of intestinal contents occurs through a rupture in the spermatic sac, usually just external to the deep inguinal ring. The parietal peritoneum is ruptured, so prolapsed intestine comes to lie cranial to the spermatic sac in the loose fascia without a peritoneal sac.

The <u>foal</u> with this condition may have a history like that for a congenital hernia or there may be a history of the foal having been kicked or otherwise injured immediately prior to the development of scrotal swelling. It may be difficult to disinguish rupture from hernia in the early stages though the swelling does not resolve and the contents are not usually totally reducible. Moreover the foal may not grow as well as expected or actually deteriorate. Colic, except briefly after the original traumatic incident, is unusual though discomfort is more marked than with congenital hernia. Any apparently congenital hernia which does <u>NOT</u> progress towards spontaneous resolution should be investigated surgically.

Inguinal rupture is also seen in <u>stallions</u> following traumatic incidents similar to those described above under mature stallion, so it presumably has a similar aetiology. The clinical picture is similar to that shown by the mature stallion with inguinal hernia as the intestine invariably becomes strangulated. The two conditions may only be distinguished at surgery, unless the intestines can actually be seen below the skin.

IV   RUPTURE OF THE PREPUBIC TENDON

This condition occurs in mares, usually in late pregnancy or in early lactation. The prepubic tendon appears to degenerate, separating from its attachments to the pelvis and the onset of the condition is gradual. A swelling appears just cranial to the udder and, at first, may be mistaken for parturient oedema. The animal has increasing difficulty in moving, may have difficulty in delivery of the young and will begin to lose condition. It seems likely that older textbooks called this condition (incorrectly) a femoral hernia.

It is not possible to repair the rupture as the tendon is in shreds and is very weak. An affected animal should only be kept alive if it is humane to do so and for the sake of the young. The use of bands of strong towelling to other material round the abdomen may afford temporary relief but care must be taken to avoid too much pressure being applied over too little an area of the back.

C. SURGICAL TECHNIQUES

As almost all inguinal hernias in animals involve the vaginal ring through which the spermatic vessels pass, it is not usually possible to save the testis on the affected side. Post-operative swelling in the inguinal canal occludes testicular blood supply and produces testicular atrophy. Post-operative reaction in the scrotal and inguinal areas may also produce infertility in the other testis and this may be temporary or permanent. Moreover, congenital inguinal hernia is known to have an hereditary basis and it is questionable whether it is ethically correct to attempt to retain breeding potential in any affected animal. It is unethical to correct such a defect with the intention of deceiving a potential buyer of the true nature of the animal. In any case, in most instances the animal is presented for castration and the presence of an inguinal hernia is a complication to that operation rather than being the primary reason for the operation. The techniques described below, therefore, all involve castration on the affected side.

I THE PIG

The economics of performing hernia repair in the pig require careful discussion with the client as the operation may cost more than the pig is worth.

Two techniques are available:

(a) 'Open Method'. No anaesthetic is required in the small pig, whilst in the larger pig anaesthesia normally needed for castration will be required. The testis on the normal side is removed as in a normal castration. The contralateral testis (that of the affected side) is tensed against the scrotal septum and a small hole made through this from castration side to affected side. This hole must be just big enough for the testis to be pulled through and not large enough to allow bowel through. After delivery of the testis through the hole, the testis is removed by any acceptable method. This technique has the disadvantage that it is not easy to make the hole in the scrotal septum of just the right size and in a proportion of cases the intestines will follow the testis out through it.

(b) 'Closed Method'. It can be performed on a small pig with physical restraint and infiltration of local anaeshtetic along the line of skin incision. In the larger pig general anaesthesia is required. This technique is illustrated in Figures 3:7-3:10. The method gives complete security against intestinal prolapse if properly performed and, in experienced hands, takes very little time. Occasionally adhesions prevent the intestines being returned to the abdominal cavity by twisting up of the spermatic sac. If the value of the pig is such that a longer operation is economic, the sac may be opened and the adhesions broken down carefully to allow intestines to be returned to the abdominal cavity.

II THE RAM AND THE BULL

Techniques have been described for repairing acquired inguinal hernias in the ruminants and retaining the testis on the affected side. Usually omentum is used to plug the vaginal ring. However, there appears to be no evidence that such testes retain or recover normal fertility and the only justification for attempting such corrective surgery is cosmetic. As such action may deceive a buyer of the true nature of the animal it is not to be recommended. The hernia should, therefore, be reduced as described for the pig.

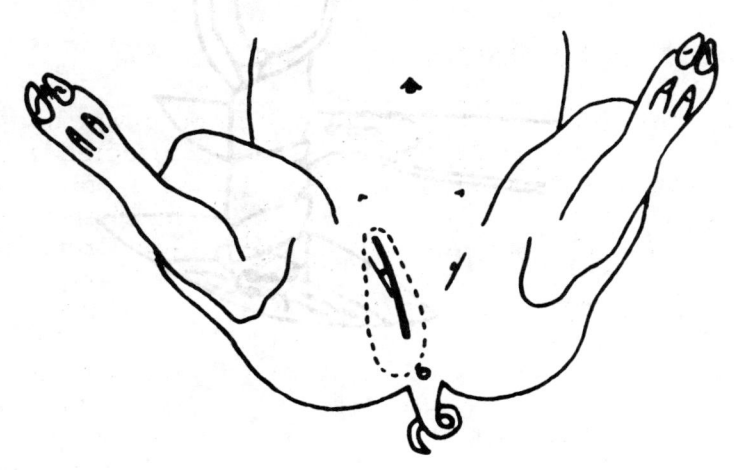

Fig. 3:7

The dotted line shows the extent of the hernia, in this case a right-sided one. An incision is made through the skin along the line indicated by a solid bar over the superficial inguinal ring.

Fig. 3:8

Working through the incision, the spermatic sac is dissected out. There is usually a layer of relatively thick fascia to be cut through before the more transparent but tough fascia of the spermatic sac is reached. The sac should not be incised. It will be found to be strongly held in place distally by the scrotal ligament which will need to be snapped.

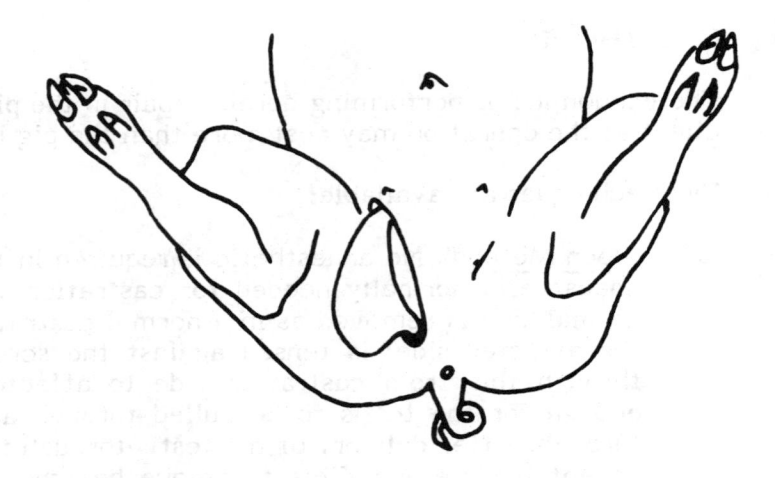

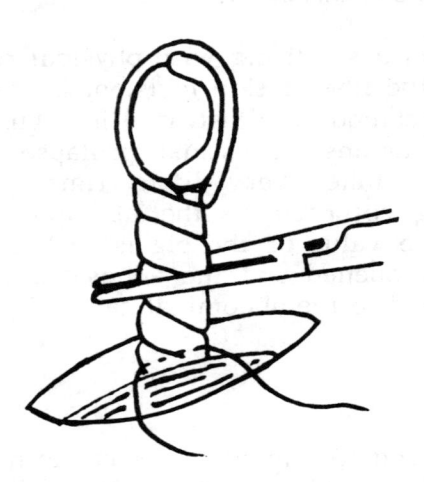

Fig. 3:9

The freed sac is twisted up to return the intestines to the abdomen. A pair of haemostats is used to keep the sac twisted whilst a transfixing ligature is applied as high up the sac as possible. It is not necessary to anchor the sac to the nearby abdominal wall. The ligature should be of strong absorbable suture material.

Fig. 3:10
The excess spermatic sac is removed and the skin incision sutured with absorbable suture material.

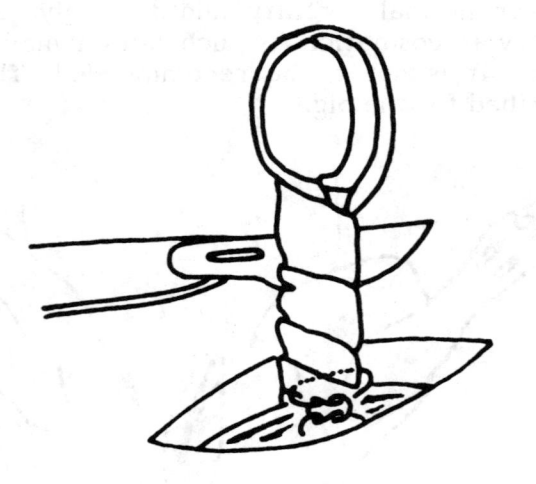

III THE HORSE

The variety of ages at which inguinal hernia occurs and the many different ways in which it may be presented in the horse mean that the approach to each case is somewhat different and is dictated by the circumstances. The simple techniques described for the pig cannot be safety applied to the horse. The pig only requires such repair as will enable it to be fattened for slaughter during a short and lazy life. The horse, however, requires such repair as shall allow it to rise safety to its feet after anaesthesia and to run, jump, buck and kick for many years afterwards. In addition, the inguinal canal of the horse is more difficult to gain access to, lying, when the horse is on its back, at the bottom of a furrow formed by the abdominal wall and the medial muscles of the thigh. All cases require the patient to be under general anaesthesia in dorsal recumbency with hind legs flexed.

(a) The uncomplicated case

These are the cases where (i) the animal is fit and well, (ii) the operation is performed electively or recognised as necessary actually at castration, and (iii) the hernial sac is the vaginal tunic.

A long skin incision is made extending from the middle of the superficial inguinal ring to the scrotum. The spermatic sac is dissected out right up to the superficial ring. This will require section of the scrotal ligament which attaches vaginal tunic to scrotal skin. The plane of fascia along which this dissection proceeds most easily is that between cremasteric and external spermatic fascia and is recognised not only by the ease with which the dissection may be performed but by the naked appearance of the external cremaster muscle exposed by it. (See also page 7).

Once isolated the spermatic sac should have its intestinal contents returned to the abdomen by gentle pressure or by twisting up. Note that the spermatic sac is not opened. A ligature of 5 metric synthetic absorbable suture material is placed through the pelvic tendon of the external oblique caudo-laterally, through the twisted up sac and through the abdominal tendon cranio-medially and then tied (compare fig. 3:9). A herniorophy needle is sometimes useful for placing the sutures in the depths of the operative area though the small needles swaged on to the modern material, are much more convenient. In this way the lumen of the vaginal tunic is closed up to the deep inguinal ring (although the suture is placed in only the superficial ring) and the possibility of rehermiation through the vaginal ring eliminated. The spermatic sac is sectioned distal to the ligature. If the external ring still appears excessively large (as it will do in a congenital case), a second suture should be placed across it cranio-laterally, staggering the distance of the needle holes from the margins of the ring to avoid common lines of stress in the aphoneurosis. On no account should the caudo-medial angle be closed as it is through here that blood and lymph vessels pass, and their occlusion will lead to preputial and penile oedema.

The subcutaneous tissue is loosely sutured to occlude dead space and the skin sutured also, again with polyglycholic acid sutures.

(b) **The complicated case**

    (i) <u>Adhesions</u>. The presence of adhesions between bowel and lining of vaginal tunic may prevent simple return of intestine and <u>so</u> necessitate opening the spermatic sac. If the adhesions can be readily broken down, all is well and good. Care is always necessary to avoid damage to the intestine and occasionally resection of the adherent area and end-to-end anastomosis is required to allow replacement. The spermatic sac and superficial ring may now be closed as described above. Fortunately, adhesions are rare.

    (ii) <u>Distension of bowel preventing return</u>. This may occur in mature stallions with strangulation or in post-castration prolapses. The vaginal tunic should then be incised up to the level of the deep ring. If this fails to allow return, then the application of traction through a counter-laparotomy incision (a supra-pubic paramedian is best - see page 47), usually enables the bowel to be returned to the abdominal cavity. This solution is <u>always</u> preferable to enlargement of the superficial inguinal ring. If gross distension with fluid or gas is present this may be relieved by puncture with a large gauge needle through a previously laid but not tied purse string suture. Once the distension has been relieved the needle is removed and the purse-string suture pulled tight to close the hole. Once the bowel has been returned to the abdomen, the spermatic sac and superficial ring should be closed as described previously.

    (iii) <u>Traumatized intestine</u>. This is most likely to occur in post-castration prolapses although strangulation in a mature stallion may have been so severe or so prolonged as to cause serious damage. All prolapsed intestine should be cleaned with warm saline. All non-viable intestine must be removed and anastomosis of healthy gut performed. In a case of post-castration prolapse it is unwise to occlude dead space in the inguinal area or to suture the scrotal skin completely as the operative site is grossly infected. Instead the area should be packed with gauze plus antiseptic or a Penrose drain may be fitted and the scrotal skin partially sutured. Convalescence, if the animal survives the surgery, may be prolonged. Almost invariably the animal is entering or has entered a state of shock and careful and continued treatment and evaluation of the animal's condition is necessary if its life is to be saved. Even so, the prognosis remains guarded and in a case where the prolapsed intestine has been outside the scrotal wound for several hours, it is unlikely that the animal will survive.

    (iv) <u>Inguinal rupture</u>. As has been stated, this condition is only likely to be differentiated from an inguinal hernia at operation when intestines will be found lying subcutaneously. It is easier to deal with such a case if the animal is first castrated on the affected side. The intestines must then be returned to the abdominal cavity by gentle massage or traction from a counter-laparotomy incision and the tear sutured as well as possible with synthetic absorbable suture material. In foals in which the rupture has been present for some time, this dissection may be difficult if not impossible and in such a case the superficial ring should be closed as well as possible.

(c) **Post-operative care**

Anti-tetanus serum or a tetanus toxoid booster dose should always be administered. A course of antibiotic therapy for four days is also indicated as the operation is somewhat prolonged and the tissues thus exposed to infection. The first dose of antibiotic should always be given prior to the commencement of surgery. Local antibiotics are contraindicated.

In the horse which has had strangulated or prolapsed intestines a careful watch for the development of shock is necessary and, as already stated, in any case which required intestinal surgery, shock is a likely sequelae. Supportive or combative therapy may be indicated during surgery.

Oedema of the inguinal area is to be expected. Indeed, it is largely the development of oedema and its resolution by fibrosis which seals off the neck of the spermatic sac and so completes the repair which the surgeon initiated. However, the oedema may spread to penis and prepuce and support for these structures may be required (p.77). Exercise may help to dissipate oedema but it must be noted that walking is sometimes a painful procedure in the first few days and forced exercise may be necessary.

Reherniation or breakdown are unlikely except as a result of poor surgical technique or in the event of gross sepsis.

D. REFERENCES

Ashdown, R.R. (1963) The anatomy of the inguinal canal in the domesticated mammals. Veterinary Record 75 : 1345-1351.

Day, F.T. (1966) In Symposium on Castration. Proceedings of the British Equine Veterinary Association for 1966.

Hutchins, D.R. and Rawlinson, R.J. (1972) Eventration as a sequel to castration in the horse. Australian Veterinary Journal 48 : 288-291.

Meredith, J.A. (1891) A report upon seven successive cases of hernia. Veterinary Record 4 : 154-156, 170-171.

Mikami, H. and Fredeen, H.T. (1979) A genetic study of cryptorchidism and scrotal hernia in pigs. Canadian Journal of Genetics and Cytology 21 : 9-19

O'Connor, J.P. (1971) Rectal examination of the cryptorchid horse. Irish Veterinary Journal 25 : 129-131.

Orr, A.E. (1956) Inguinal hernia in sheep. Veterinary Record 68 : 2-4.

Schneider, R.K., Milne, R.W. and Kohn, C.W. (1982). Acquired inguinal hernia in the horse: a review of 27 cases. Journal of the American Veterinary Medical Association 180 : 317-320.

Warwick, B.L. (1926) A study of hernia in swine. Wisconsin Agricultural Station Research Bulletin No. 69.

Wensing, C.J.G. and Colenbrander, B. (1973) Cryptorchidism and inguinal hernia. Proceedings Koniklijke Nederlandse Akademie Wetenschappen Series C 76 (5) : 489-494.

Wright, J.G. (1963) The surgery of the inguinal canal in animals. Veterinary Record 75 : 1352-1367.

Chapter Four
THE EQUINE PENIS AND PREPUCE

A. ANATOMY

The equine penis, like that of other species consists of two erectile tissues. The <u>corpus spongiosum penis</u> surrounds the urethra from its emergence at the pelvic outlet. It has two enlargements, one caudally known as the bulb of the penis of the horse and has a characteristic caudal extension over the corpus cavernosum (Figs 4:1 and 2). The <u>corpus cavernosum penis</u> has its origins in two crura near the ischial arch on either side of the pelvic outlet. These two halves fuse distally to form a single body which extends to the tip of the penis. Each crus is overlain by a powerful ischio-cavernosus muscle and the corpus spongiosum is overlain throughout its length by the bulbospongiosus muscles whose fibres are transversely orientated. The retractor penis muscles of the horse are paired but close. They decussate through the bulbospongiosus muscles to insert on the tunica albuginea at the middle of the penis. They do not, therefore, appear in Fig. 4:3, which is a section further distally.

The internal pudendal artery supplies the root of the penis, especially the bulb (= caudal enlargment of corpus spongiosum). A branch of the obturator artery and a branch of the external pudendal artery on each side join the dorsal artery of the penis. Branches of this artery supply the bulb of the penis and the corpus cavernosum penis at the ischial arch and further small branches supply the external region of the tunica albuginea. The veins, which form a rich plexus dorsally and on either side of the penis, drain into veins corresponding to the arteries, except that the external pudendal vein does not pass through the inguinal canal (see p.69). As in ruminants, the blood supply to and the venous drainage from the corpus cavernosum is entirely at the root and the action of the ischiocavernosus muscles is initially to occlude venous drainage (producing turgidity) and then to pump blood distally to achieve erection.

The equine penis is of the musculo-cavernous type as its cavernous spaces are large relative to the trabeculae. Associated with the trabecular network of the corpus cavernosum penis are thick bundles of smooth muscle fibres. These bundles are orientated longitudinally and are normally in a state of tonic contraction, thus holding the penis in its prepuce as shown in Fig. 4:1. If the tonus in the muscle bundles of the trabeculae falls, then the penis falls out of the prepuce. This happens normally at micturition - the penis may be said to be <u>flaccid</u>. It can also happen under a variety of pathological states (see below page 91) and this makes the penis of the gelding more susceptible to trauma than in the castrates of other species of animals. This extrusion occurs in two stages due to the complex integumentary folding. Initially the free end of the penis is extruded within the fold of integument known as the preputial fold (Fig. 4:1). Later, the free end of the penis is extruded through the preputial ring which marks the boundary between the inner and outer layers of the preputial fold.

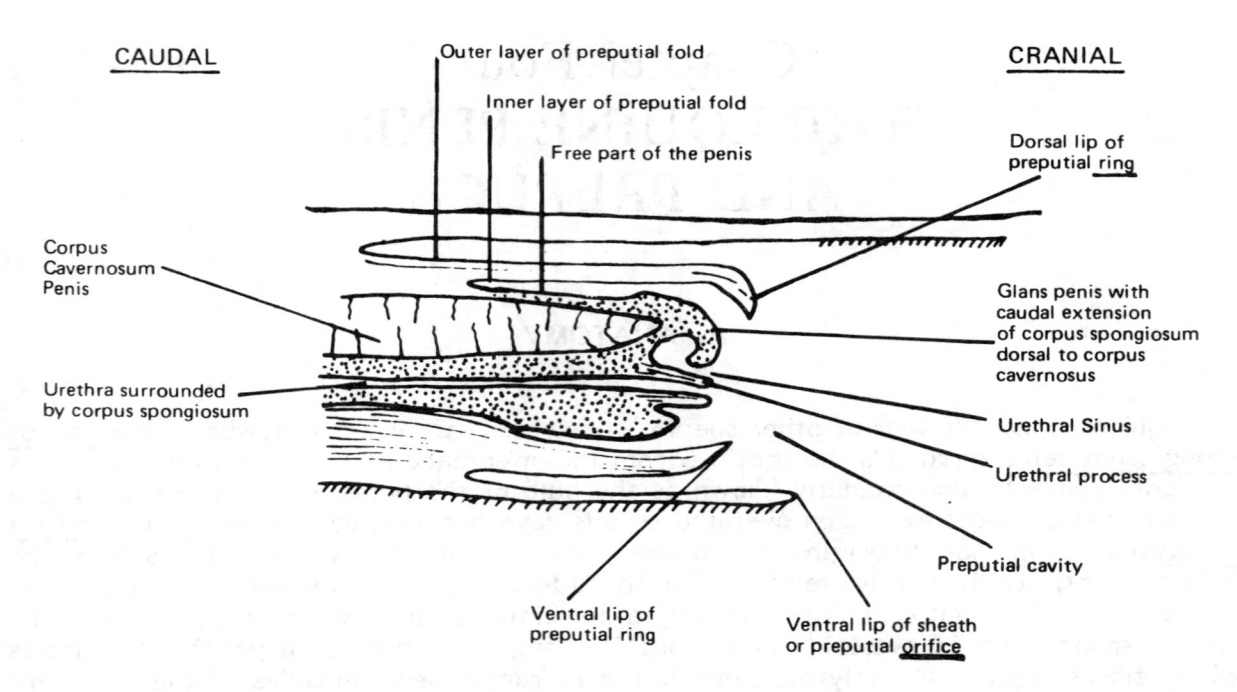

Fig. 4:1 Longitudinal Section through equine penis in the mid line.

The tonus of the muscle bundles also falls during sexual stimulation. Initially the cavernous spaces become filled with blood at arterial pressure and the flaccid penis stiffens. Such a penis is referred to hereafter as turgid. A turgid penis may develop during Immobilon neuroleptanalgesia. A turgid penis can be generated by the horse in the absence of mares and sometimes this is associated with partial masturbation during which the horse flicks his penis against his abdominal wall. As sexual excitement develops, the blood pressure within the corpus rises due to activity of the ischio-cavernosus muscles which overlie the crura. Pressures within the corpus cavernosum penis may reach 50 x arterial pressure (Beckett et al, 1973). The increased amount of blood and the pressure transform the turgid penis into an engorged erect one, such as is shown in Fig. 4:2, the degree of engorgement being limited by the tunica albuginea (Fig. 4:3). The fully erect penis characteristically has an enlarged distal tip forming a corona glandis.

The integumentary covering of the penis consists of two distinct parts (Fig. 4:2). That overlying the outer layer of the preputial fold is a stratified epidermis with sweat and sebaceous glands. The secretions of the latter combine with epithelial debris to produce the dirty grey, evil smelling, somewhat cheesy smegma which accummulates in the depths of the preputial recess. (Strangely, donkeys do not accummulate smegma). The integument overlying the inner layer of the preputial fold is overlain by a non-glandular stratified epithelium with numerous deep rete pegs. The keratinised tissue forms flakes which often lie loosely attached to the underlying epithelium in horses whose penis' are not interfered with. In the lighter coloured horses it often has pink patches. The epithelium over the corona glandis is specialised into conical papillae.

The urethra does not open simply at the end of the penis but is surrounded by a depression, the urethral sinus, into which it protrudes as the urethral process. The urethral sinus has a dorsal diverticulum.

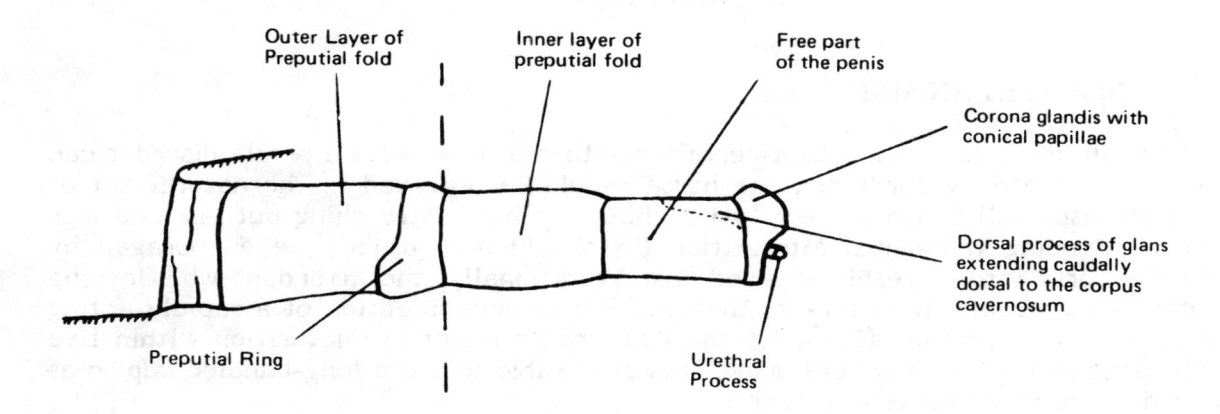

Fig. 4:2 Extended Equine Penis.

The penis is supplied by nerves from the 3rd and 4th sacral nerves via the pudendal nerve and its chief branch, the dorsal nerve of the penis, and also via the pelvic nerve (nervus erigens). Sympathetic innervation via the pelvic plexus is from the hypogastric nerve.

The lymphatic drainage from the penis and prepuce is to the superficial inguinal lymph nodes (palpable just cranial to the inguinal ring) and the deep inguinal lymph nodes (near the femoral ring) and thence to the medial iliac nodes round the terminal portion of the aorta (palpable per rectum). It is important to recognise that much of the lymphatic and venous drainage is through the inguinal region. Conditions, such as post-castration infections, which impede this drainage, almost invariably cause swelling of the penis and prepuce.

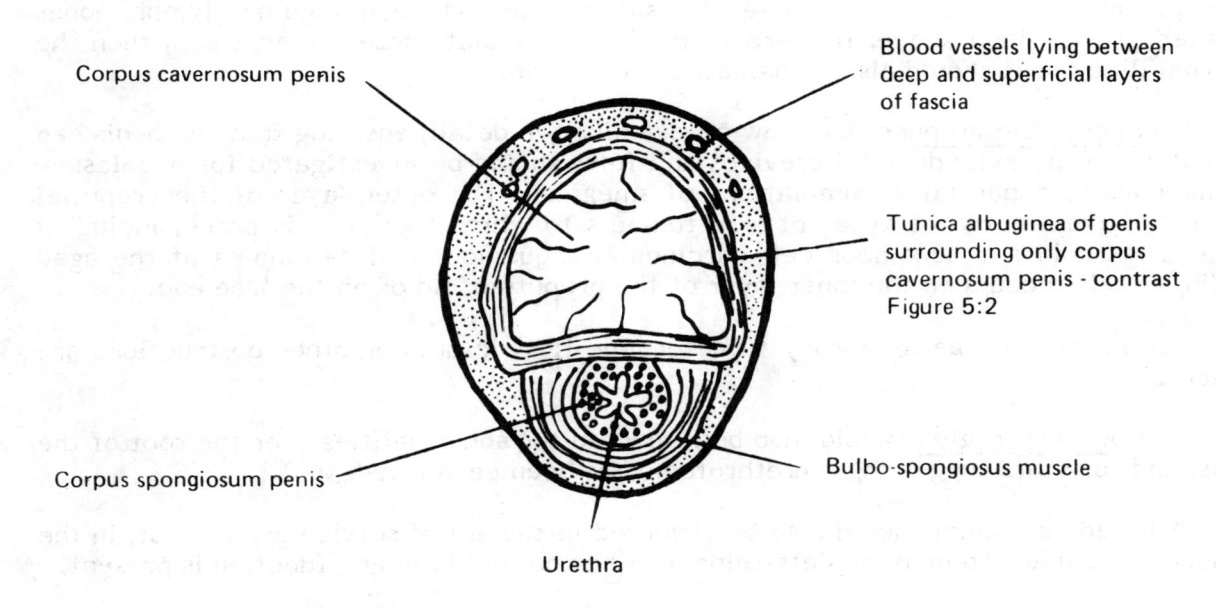

Fig. 4:3 Cross-section through the Extended Equine Penis.

B. EXAMINATION

I THE DISEASED ANIMAL

It is of the utmost importance to ascertain whether a horse with a penile disorder can or cannot micturate. Accordingly, the horse should be observed in the natural act of micturition, especially with a view to watching the horse stretching out his hind legs and then extruding the penis. Micturition ("Staling") may usually be encouraged by putting the horse into a freshly strawed box. Traditionally, the attendant whistles and shake the straw about. If this fails, then the intravenous injection of a rapidly acting diuretic, such as frusemide (0.5 - 1.0 mg/Kg), should result in micturition within five minutes. (If it is desired to collect urine, it is preferable to use a long-handled cup so as to avoid disturbing the horse too much).

A cause of difficult micturition which should always be considered is accumulation of hardened epithelial debris within the urethral sinus - this accumulation is usually readily palpable and should be carefully removed. If micturition is not possible, careful catheterisation should be attempted immediately to relieve the pressure in the bladder. Should this prove difficult, dangerous or impossible because of stricture, calculus or oedema, then a sub-ischial urethrotomy should be performed.

II DETAILED PHYSICAL EXAMINATION

(a) The prepuce should be examined by looking and by palpation, for injuries, swellings, accumulations of secretions or other space-occupying lesions. Papillomas and other skin tumours are not uncommon on the sheath.

(b) The inguinal region and scrotum should be examined for scars of castration (see p.43), for discharging fistulas (see p.19) indicating suture abscess or scirrhous cord, for swelling indicating other post-castration complications and for testes. In addition, an attempt should be made to palpate the superficial and deep inguinal lymph nodes situated as described above. If there is any evidence that these are enlarged, then the internal iliac nodes should also be palpated per rectum.

(c) The penis and prepuce may now be examined in detail, ensuring that the penis can normally be fully extended. All crevices and folds should be investigated for neoplasms, granulomas or abnormal accumulations of smegma. The outer layer of the preputial fold may have any of the types of skin tumours to which the horse is prone, including fibroma, sarcoid and squamous cell carcinoma. Squamous cell carcinoma of the aged gelding usually occurs on the inner layer of the preputial fold or on the free end.

Catheterisation may be necessary to determine if strictures or other obstructions are present.

(d) The perineal region should also be inspected for abnormalities over the root of the penis, and for scars suggesting a urethrotomy has been performed.

(e) A breeding animal should also be observed in the act of service or, at least, in the presence of a mare on heat, to determine whether normal breeding function is present.

C. CLINICAL CONDITIONS

I CONGENITAL CONDITIONS

These are generally rare but the following have been recorded (Vaughan, 1972):

(a) congenitally short penis

(b) dysfunction, including aplasia of erectile tissue of glans

(c) preputial aplasia

(d) hypospadias, in which the urethra opens on the ventral surface of the penis or in the perineal region

(e) intersex horses - variety of findings but there is often an enlarged penis-like clitoris.

II SPECIFIC INFECTIOUS DISEASES

(a) <u>Equine coital exanthema</u> (= 'Horse Pox')

This is a contagious disease caused by an equine herpes virus antigenically distinct from the type 1 (equine rhinopneumonitis virus) and from the types 2, 3, etc. (equine cytomegalo viruses). Generally speaking, the disease is transmitted venereally but mechanical transmission by muzzling by other mares or by teaser stallions or perhaps by flies or even veterinary surgeons is not impossible. The incubation period varies from 3-6 days, but in many instances is only 12-24 hours. The disease is characterised by the development of papules, with surrounding hyperaemia and oedema, pustules and eventually ulcers. In the stallion the lesions appear especially on the free end of the penis and in the mare inside the vulva and on the surrounding perineal skin. Lesions may also appear in the mouth. The density or extent of lesions may occasionally be so great that the lesions coalesce, sometimes resulting in large ulcerated areas. When healing takes place, smooth white (depigmented) areas remain.

The disease is widespread throughout the world (including the U.K.) but may not be noticed unless there is routine veterinary inspection prior to mating or the stallion is involved. Although the secondary infection may result in vaginal and preputial discharge, this is generally without effect on the fertility of mare or stallion. However, discomfort may be shown in the form of apparent sexual excitement, frequent attempts at staling and rubbing of the tail. Occasionally, however, severe secondary infection in the stallion may result in loss of libido and it has been suggested that the virus can induce premature luteolysis by setting up a mild endometritis.

Treatment of affected animals should be aimed at controlling secondary infection by washing affected areas daily with a weak antiseptic solution - Vandeplassche (personal communication) prefers a suspension of sulphonamides in oleum jecoris asselli and Balsam of Peru sprayed onto the erect penis. In severe and neglected cases in which the animal is systemically ill, systemic antibiotic therapy may be justified.

Control is difficult. Although it is probably that a proportion of affected animals remain as asymptomatic carriers, normally the animal ceases to shed virus once the lesions have healed (3-4 weeks, occasionally less). Affected animals, therefore, should be isolated and not allowed to mate so as to reduce the spread of the disease. Because of the high possibility of mechanical transfer, infected animals should be handled with disposable gloves.

(b) Dourine

This is an infectious disease of Equidae caused by Trypanosoma equiperdum characterised by genital, cutaneous and nervous manifestations.

- (i) Distribution. Dourine is prevalent in parts of the Middle East, the Balkans, North Africa and South America and in small pockets of South Africa. It was eradicated from Canada in 1919, but persisted until 1949 in wild horses in the U.S.A. It does not occur in Northern and Western Europe or in the Far East, Australia and New Zealand.

- (ii) Transmission. Dourine is transmitted venereally but not all matings are infective. Although T. equiperdum cannot survive off the host for long, it can also be transmitted by equipment used for artificial insemination. Foals can be infected through the conjunctiva by vaginal discharge from the mare.

- (iii) Pathogenesis. The primary lesion occurs at the site of infection where the parasite multiplies before spreading haemotogenously to other sites. The organism causes vascular injury which results in oedema and sometimes ulceration and also causes degeneration of peripheral nerves with paralysis and muscle wasting.

- (iv) Clinical signs. The course of the disease is variable. In Europe it is commonly acute with a succession of crops of urticarial-like plaques 2-5 cm in diameter developing at intervals with death ensuing in a few weeks. Elsewhere, however, the disease is more chronic. In the stallion the initial sign is oedema of the penis, prepuce, scrotum and surrounding skin and this oedema may extend along the ventral abdomen. Paraphimosis may result from this oedema. The inguinal lymph nodes are swollen and there may be a mucopurulent urethral discharge. In mares, oedema is first seen in the vulva but spreads to the perineum, udder and ventral abdominal wall. There is usually a profuse vaginal discharge with hyperaemia of the vagina. In both sexes, nervous signs eventually develop with stiffness and weakness, although this may take several years. Emasiation is common and can be severe.

- (v) Treatment. Affected animals may recover if treated with trypanocidal drugs in the early stages of the disease, but treatment of any sort is of little use in the later stages. Treatment should not be attempted if eradication of the disease is planned.

- (iv) Control. As T. equiperdum is primarily a tissue parasite, rather than a haematogenous one, it can only be detected in the blood during the short period of haematogenous spread. Accordingly, infected animals cannot be detected by examination of blood smears as is the case with most other trypanosomal diseases. Diagnosis of infection can, however, be made on the basis of a complement fixation text (CFT) which becomes positive about 3 weeks after the initial infection. By combining diagnosis by blood test using the CFT with a policy of slaughtering positive animals the disease can be eradicated.

III TRAUMATIC DISORDERS AND PARALYSIS

(a) Aetiology

Stallions may suffer abrasions, contusions and/or lacerations of the penis during service. Stallions may also suffer trauma to the penis from an ill-fitting ring. Both stallions and geldings may have their penis injured by kicks during fighting or, more maliciously, by the penis being whipped. Such injuries usually result in haemorrhage from the rich plexus of veins outside the tunica albuginea. Rupture of tunica itself, such as occurs in bulls (p.103), is probably extremely rare in the stallion.

Untreated oedema of the sub-integumentary layers of the penis may resolve by fibrosis. Such a lesion may prevent the horse either extruding the free end of the penis or withdrawing the free end back inside the preputial fold because it prevents rolling and unrolling of the preputial fold.

In areas of the world where Habronema muscae or Callitroga spp. exist, their larvae are often found in the moist inviting recesses of the prepuce. Their migrations and encystment are often accompanied by granuloma formation infiltrated with eosinophils. Although responsive to appropriate topical and systemic treatment, healing is by scar tissue which may prevent subsequent normal retraction of the penis.

Prolapse of the penis may occur in a number of different situations.

(i) Sheer physical exhaustion was often reported as a common cause of penile prolapse in the heyday of the horse.

(ii) Occasionally prolapse may be seen as a sequel to rabies or to other central nervous system disorders.

(iii) The administration of phenothiazine derivatives is followed by penile prolapse and very occasionally this prolapse is irreversible. This seems more likely to arise if excessive doses are employed, the animal is an entire and the injection is given intravenously (see Wheat 1966 and Jones, 1966). This prolapse has been reported recently to follow the use of the neuroleptanalgesic 'Immobilon' which contains acepromazine (see Pearson and Weaver, 1978) and in such cases is often associated with turgidity (not erection) of the penis.

(iv) The author has also observed penile prolapse as the most dramatic sign of general systemic disease, in particular malabsorption syndrome (Simmons, Cox, Edwards, Neal and Urquhart, 1985). In such cases the penis is completely prolapsed and paralysed and is also usually grossly oedematous. The oedema may spread along the ventral abdomen. Analysis of blood proteins reveals no consistent abnormality. Wherever prolapse of the penis is seen without any obvious cause, the presence of intercurrent disease should always be considered.

(v) Gross swelling, perhaps associated with infection, in the inguinal area after castration or bad dissection during the search for a cryptorchid testis may interfere with lymph drainage from the penis through the inguinal canal. As a result of this lymphatic blockage, the penis and prepuce may swell and prolapse through the preputial orifice.

(b) <u>Pathology</u>

Figure 4:4 is a schematic representation of the sequence of pathological events in a prolapsed/traumatized penis. Causes are given peripherally and sequelae centrally. It is important to realise that in a prolapsed damaged penis, normal lymph drainage is impeded, not only because of pathological processes occluding lymph vessels, but also because the forces of gravity hinder fluid movement. A prolapsed penis, therefore, swells because of fluid accumulation and prolapses further, thus initiating the cycle shown in Fig. 4:4. In a neglected case, the integument dries (particularly that covering the inner layer of the preputial fold and free end) and its surface cracks, allowing bacteria to enter and set up a classic inflammatory response with oedema, thus completing the cycle of events. The result of injury is, therefore, almost invariably balanoposthitis with paraphimosis. If the paraphimosis is restricted to the preputial fold and the free end itself remains entirely within the preputial fold, then urine scald can be a serious complication.

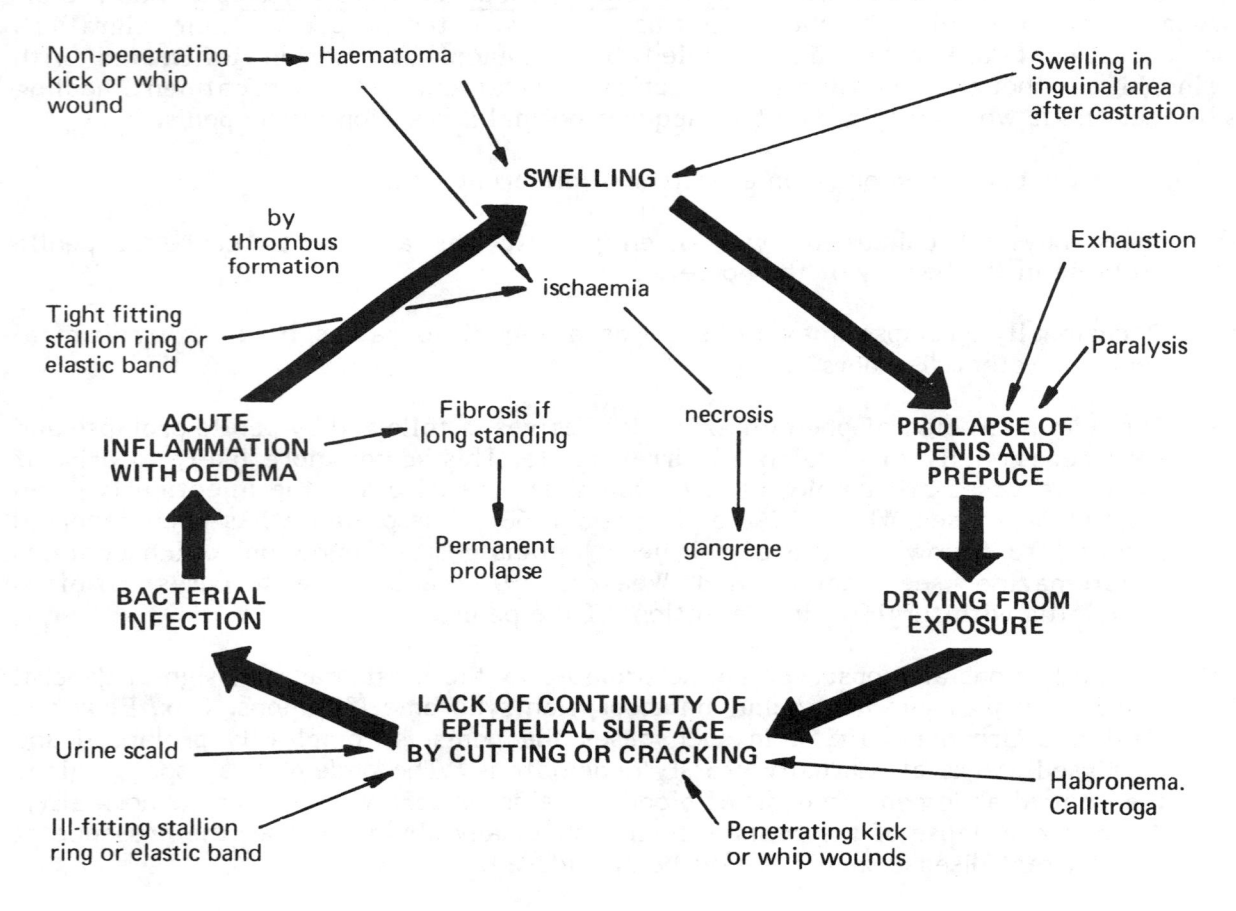

Figure 4:4 Schematic Representation of Pathological Events in a Damaged Equine Penis.

(c) Treatment

(i) Non-operative Procedures

The earlier treatment is initiated the more rapid will be the response and the less time healing will require. Moreover, it is likely that success or failure in the majority of cases of traumatized/prolapsed penis will depend upon the skill with which nursing care is employed. The aim of treatment is to break the vicious cycle shown in Fig. 4:4 by providing support for the penis by actively reducing the oedema and by protecting the surface from further damage.

1. Reduction in Swelling. In the early case, the application of pressure will prevent the development of excessive swelling or will reduce swelling that has already developed. Pneumatic bandages have been advocated by Vaughan (1972) and elastic crepe bandages by Pascoe (personal communciation). If facilities for compression are not available, cold water packs or hosing with cold water are beneficial. Exercise, enforced if necessary, will also help to dissipate oedema. Lucke and Sansom (1979) note that when the condition appears under general anaesthesia, it is best dealt with whilst the animal is still anaesthetised.

In later stages, warm moist applications may help to dissipate oedema. Systemic anti-inflammatory drugs and diuretics may also be useful, but the latter will be effective only if the animal is allowed restricted access to water. In general, it is better for the horse to be in a field where it can potter about than confined in a loose box. The penis should also be supported (see 2 below).

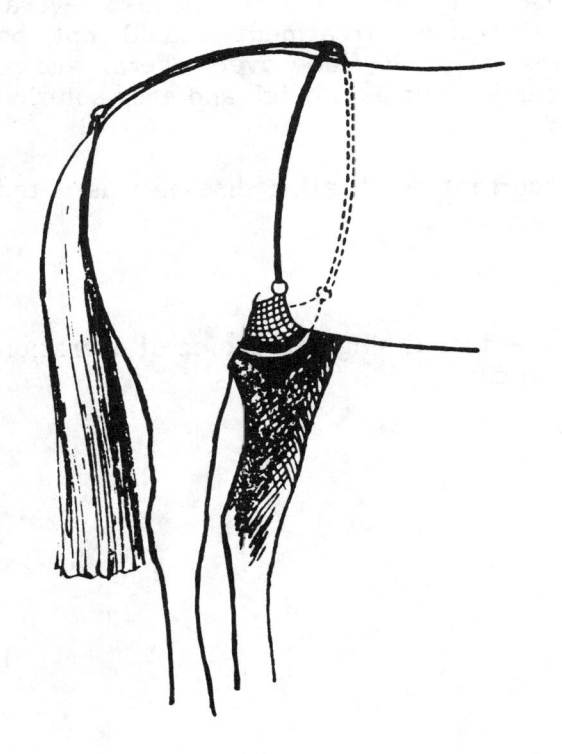

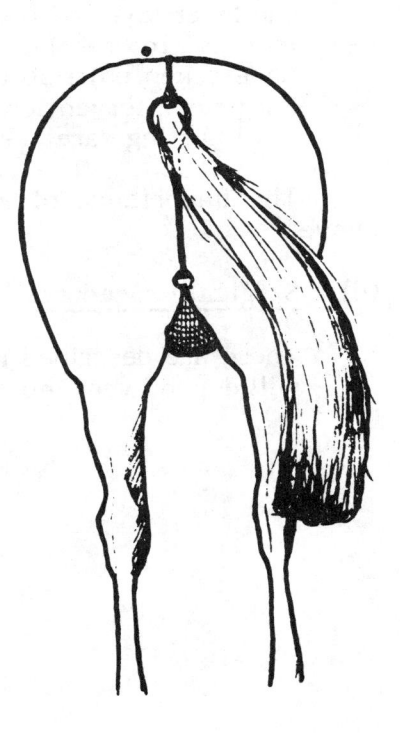

Fig. 4:5 Suspension for a prolapsed penis with crupper and surcingle.

2. Support. In a case of gross prolapse, support is essential and may be provided by an open mesh nylon net, e.g. old pantie hose, supported by a surcingle and crupper made of rubber tubing. The mesh allows urine through but urine scald still needs preventing by emollient. An alternative is suitably protected plastic guttering of appropriate length - all corners must be smoothed off. The guttering provides a convenient channel for the penis to lie in. The importance of support in reducing swelling cannot be over-emphasised.

3. Protection. The entire surface of the penis should be protected by the application of an ointment such as petroleum jelly which will prevent drying and cracking from exposure or from urine scald. If open wounds are present then an antiseptic ointment, to help control infection, is preferable until healing is well under way.

4. Specific Lesions. Granuloma from insect larvae require specific therapy. If there is inguinal lymph blockage following castration, steps will need to be taken to deal with this by opening the scrotal incision and ensuring adequate drainage (see p. 17ff).

5. Prognosis. Care may be required for extended periods of time. The support requires changing, the penis requires cleaning and a new application of ointment are needed daily. Large areas of denuded surface can, and do, epithelialize rapidly, especially from the rete pegs of the free end and the inner layer of the preputial fold. Nevertheless, the horse may not retract his penis, although able to do so, until epithelialization is complete. Moreover, slight swelling and fibrosis, particularly on the dorsal surface of the inner layer of the preputial fold (see Figs. 4:1 and 4:3) may take several months to resolve satisfactorily. Operative treatment should not be undertaken until absolutely no progress at all is made over several weeks. Surgical intervention should not be carried out as a quick and easy solution until nursing care has obviously failed.

The importance of eliminating intercurrent debilitating disease was noted above.

(ii) Surgical Procedures

These are described in Section D. The actual timing and choice of operation will depend very much on the individual case.

IV NEOPLASTIC CONDITIONS

(a) Squamous Cell Carcinoma

This cauliflower-like growth occurs on the penis, especially the free end and inner layer of the preputial fold, of aged geldings. Similar lesions occur on the clitoris of aged mares. Keratinization may be present and the lesion is often ulcerated and fungating, and may, therefore, need to be distinguished from Habronema granulomas.

Gelding smegma is one of the few natural secretions known to be carcinogenic (Plaut and Kohn-Speyer, 1947) but the tumour grows surprisingly slowly for a carcinoma. Local metastasis to other areas of the free end and preputial fold may occur. In advanced cases, in which the penis cannot usually be extruded, systemic metastasis may occur. However, in a case which has only been growing for some months, the local lymphadenitis seen is usually due to secondary infection. It is unwise, therefore, to recommend euthanasia solely because of local lymph node enlargement. Treatment is by reefing or by amputation (p.84), if possible, but if it is impossible to exteriorize the penis, euthanasia is indicated.

(b) **Other neoplasms** include sarcoids, squamous papilloma, melanoma and haemangioma. All the possible types and combinations of types of skin tumours that can occur in horses can occur on the outer layer of the preputial fold and its opposing surface. Occasionally the growths may be so extensive as to justify a pre-scrotal urethrostomy and excision of the prepuce.

V MISCELLANEOUS CONDITIONS

(a) Rupture of the preputial fascia

One case has been recorded by Swanstrom and Krahwinkel (1974). Rupture of the suspensory fascia had resulted in the penis falling further back into the preputial orifice producing a swelling in the prepuce just cranial to the scrotum. The horse was unable to extrude the penis and so urinated into the preputial cavity, producing severe urine scald. Repair of the ruptured fascia was achieved at surgery and the condition resolved.

(b) Penile cysts

One case has been seen by the author where a large number of cysts of varying size occurred subcutaneously in the inguinal region and on the caudal part of the penis. These first became apparent when the animal was two years old and became progressively larger, eventually resulting in prolapse of the penis. Resection of some of the larger cysts resulted in improvement.

(c) **Haemospermia due to urethritis**

A group at Colorado State University (see Voss and Pickett, 1975) have described their investigations of persisting haemospermia in the stallion. Although semen quality, etc. were normal, affected animals occasionally showed evidence of pain on ejaculation and were invariably infertile.

Most cases were due to a bacterial urethritis with the most commonly affected area being that adjacent to the openings of the ejaculatory ducts. The syndrome appeared to be more common in the most heavily used stallions. Some cases appeared to be associated with a viral urethritis. Other, less common, causes included ill fitting stallion rings, a strongyle larva migrating through the vesicular gland and trauma to the glans from a coarse suture in a mare's vulva.

The cases were investigated by a microscopic examination of ejaculates and of cells collected from the urethra immediately after ejaculation, by direct visualisation with a urethroscope, inserted if necessary through a sub-ischial urethrotomy incision, and by contrast radiography involving injection of contrast medium and air.

Bacterial urethritis responded to sexual rest plus systemic antibiotic or sulphonamide therapy, supplemented by local application inserted through the urethrotomy incision. In some cases maintaining the horse on 10 gm of methenamine daily plus a urinary acidifier has resulted in control of the condition. Surgical excision of strictures has also been described.

D. SURGICAL PROCEDURES

All operations require general anaesthesia with the patient in dorsal recumbency.

I   REEFING PROCEDURE

This is indicated for the removal of neoplasias or granulomas where simple excision of the tumour is inadequate and amputation of the penis is not necessary or for removal of excess tissue formed following trauma or chronic inflammation which produces tissue in the prepuce and resultant dysfunction.

The operation is performed with the horse under general anaesthesia in dorsal recumbency. Haemostasis is achieved initially by placing tourniquet round the penis proximal to the lesion and two parallel, circumferential incisions are made through the integument distal and proximal to the lesions. The integument between the incisions is stripped away, care being taken to avoid the large branches of external pudendal artery and vein that lie just outside the tunica albuginea of the penis (Fig. 4:3). The tourniquet is released and all bleeding points identified and ligated. The integument is closed over the area by a row of absorbable sutures buried in the subcutaneous adventitia and by a second row in the integument.

A stallion should have a ring fitted for at least 2 weeks afterwards and be isolated from mares for 4 weeks.

II   PENILE RETRACTION

Penile retraction is only indicated in case of permanent paralysis or prolapse uncomplicated by excessive swelling or active inflammatory processes; the procedure is an alternative to amputation.

Stallions should be castrated well in advance and any subsequent swelling and inflammation allowed to subside. If a reefing operation is necessary, it and the penile retraction may be done together. The whole of the penis and scrotal area should be prepared for operation.

The horse is anaesthetised and placed in dorsal recumbency. The urethra is catheterized to assist in identifying the penis - a tourniquet is not required. Skin incisions, 6-10 cm long are made at the level of the scrotum parallel to the mid-line and a few cms from it. The penile fascia is bluntly dissected to free partially the penis, taking care to avoid the large branches of the pudendal vessels. The penis is pushed caudally into the incision by a surgical assistant to that the penis is fully retracted into its preputial fold and sheath. On each side a strong non-absorbable suture is inserted into the base of the preputial fold, taking care not to penetrate the integument and so enter the preputial cavity, and also taking care to avoid the urethra. These sutures are anchored through the skin 3-4 cm away from the free edge of the mid-point of the skin incision using tension buttons, quills or bandage rolls. The preputial ring should lie at the cranial edge of the skin incision. The fascia is apposed and the skin is sutured.

Any oedema which develops should be controlled by exercise, though placement of surgical drains may be effective in preventing serum accumulation. Sutures may be removed in two weeks, stallions should be kept from mares for 4 weeks to avoid undue sexual excitement.

III PENILE AMPUTATION

Amputation of the penis is indicated where there is carcinoma or permanent paralysis with irreducible enlargement rendering the penile retraction operation impossible.

Two techniques are available, one described by Williams (see Danks, 1956) and one due to Vinsot (see Shuttleworth and Smythe, 1960). Although the Williams' operation is believed to be more free of urethral stricture than the Vinsot procedure, the author has tried both procedures, finds Vinsot's easier and has had no problems of urethral stricture. It is that procedure which he now employs and recommends.

The horse is given a general anaesthetic and placed in dorsal recumbency.

The penis is fully extended and held extended by a snare, a catheter passed to ease identification of the urethra and a tourniquet applied proximally (Fig. 4:6).

A transverse incision is made about 2cm behind the intended level of amputation on the ventral surface, extending about 2cm to either side of the mid-line. From the extremities of this incision, two others are made passing caudally to join each other about 6cm behind the first. The triangular area of integument so contained is removed and a longitudinal incision made through the mid-line right down to the urethra and exposing the catheter. The edge of the urethra is sutured to the edge of the integument with carefully placed interrupted sutures of absorbable material. Ideally the sutures should pass through both incised edges of the tunica surrounding the corpus spongiosum for totally effective haemostasis but these edges cannot always be recognised.

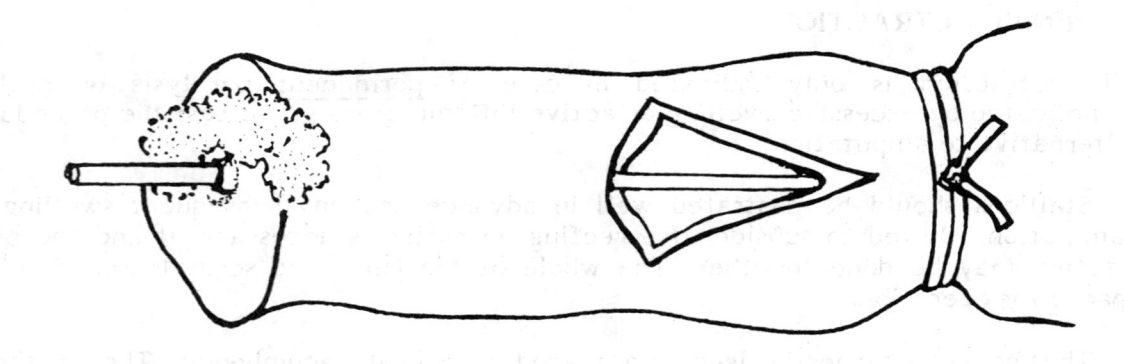

Fig. 4:6 Equine penis prepared for amputation. Fully extended. Catheter inserted. Tourniquet applied. Urethrostomy prepared with catheter visible

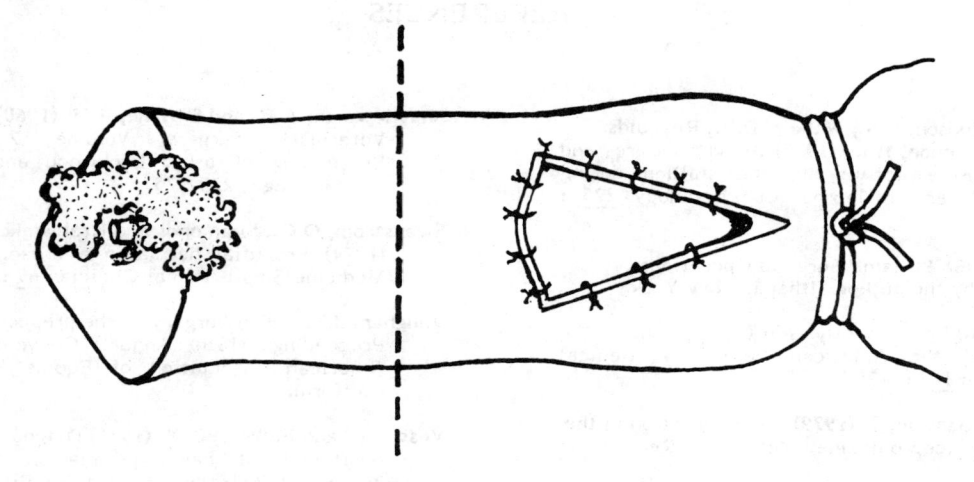

Fig. 4:7 Equine penis prepared for amputation with urethrostomy complete and catheter withdrawn. The penis is emputated along the dotted line not less than 2 cm distal to the cranial edge of the urethrostomy

The catheter is now removed and the penis amputated 2cm distal to the urethrostomy that has been created as described above (Fig. 4:7). The vascular bed of the corpus cavernosum and the large dorsal veins of the penis are now exposed. A series of mattress sutures are placed through the integument, through the tunica albuginea, through the tunica albuginea opposite and through the integument once more. (Fig. 4:8 - it helps to place all sutures before tying any.) The integument is drawn over the stump with interrupted sutures.

The tourniquet is removed and if suturing has been satisfactory there is usually no haemorrhage. The end of the penis is then left to granulate which it does satisfactorily. Daily douching for a week of the sheath with a warm antiseptic solution helps to keep the operation site clean.

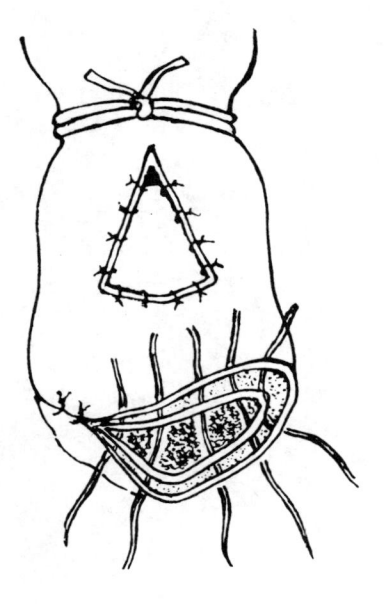

Fig. 4:8 Dealing with the end of an amputated penis. Notice the deep sutures through the integument — tunica albuginea — tunica albuginea — integument and superficial sutures through the integument only.

E. REFERENCES

Beckett, S.D., Hudson, R.S., Walker, D.F., Reynolds, T.M. and Vachon, R.I. (1973) Blood pressures and penile muscle activity in the stallion during coitus. American Journal of Physiology 225 : 1072-1075.

Danks, A.G. (1956) Williams' Surgical Operations. Published by the author. Ithaca. New York.

Jones, R.S. (1966) Penile paralysis in the stallion. Journal of the American Veterinary Medical Association 149 : 124.

Lucke, J.N. and Sansom, J. (1979) Penile erection in the horse after acepromazine. Veterinary Record 105 : 21-22.

Pearson, H. and Weaver, B.M.Q. (1978) Priapism after sedation, neuroleptanalgesia and anaesthesia. Equine Veterinary Journal 10 : 85.

Plaut, A. and Kohn-Speyer, A.C. (1947) The carcinogenic action of smegma. Science 105 : 391-392.

Simmonds, H.A., Cox, J.E. Edwards, G.B., Neal, P.A. and Urguhart, K. (1985) Paraphinosis in seven debilitated horses. Veterinary Record 116 : 126-127.

Shuttleworth, A.C. and Smythe, R.H. (1960) Clinical Veterinary Surgery Volume 2. Operative Procedure. Crosby, Lockwood and Sons Ltd., London. pages 273-275.

Swanstrom, O.G., Simmonds and Krahwinkel, D.J. (1974) Preputial hernia in a horse. Veterinary Medicine/Small Animal Clinician 69 : 870-871.

Vaughan, J.T. (1972) Surgery of the prepuce and penis. Proceedings 18th Annual Convention of the American Association of Equine Practitioners, California.

Voss, J.L. and Pickett, B.W. (1975) Diagnosis and treatment of haemospermia in the stallion. Journal of Reproduction and Fertility Supplement 23 : 151-154.

Wheat, J.D. (1966) Penile paralysis in stallions given propriopromazine. Journal of the American Veterinary Medical Association 148 : 405-406.

Chapter Five
THE BOVINE PENIS AND PREPUCE

Although this Chapter is concerned primarily with the bull, the anatomy and pathology of the penis and prepuce of the ram and boar are similar and references to this species are included where appropriate.

A. ANATOMY

I THE PENIS

The bovine penis, like that of the small ruminants and the pig is of the fibroelastic type. When fully extended it often exceeds a metre in length, appearing not unlike a 'round, elastic stick with a pointed end' (Nickel et al, 1973). In the quiescent state it forms an S-shaped curve at the level of the scrotal neck which is called the sigmoid flexure, about a quarter of the total length of the penis being involved.

Structurally there is a thick, relatively inelastic tunica albuginea forming a tube round a dense system of fibro-elastic trabeculae. Between the trabeculae are small cavernous spaces which extend the whole length of the penis and which form the corpus cavernosum penis.

The penis is anchored to the ischium by the crura (see (a) below) and to the sub-pelvic tendon by a suspensory ligament. The latter sometimes ruptures causing the penis to sag.

It is convenient to describe the penis in three parts, viz. the root, the body and the free end. (Fig. 5:1).

(a) The Root

This is formed in part by two crura which originate from the ischiatic arch between the two tuberosities. Each crus is rod-like but compressed laterally where it is covered by the ischio-cavernosus muscle.

Between the two crura runs the first part of the extra-pelvic urethra. The urethra is surrounded caudally and laterally by the cavernous spaces called the corpus spongiosum penis, to distinguish it from the corpus cavernosum penis. At the ischial arch there is an enlargement of the corpus spongiosum penis, viz. the bulb of the penis. This bulb is overlaid by the bulbo-spongiosus muscle.

At the root of the penis, the retractor penis muscles are 5cm apart but come closer together distally to lie on either side of the distal bend of the sigmoid flexure.

88

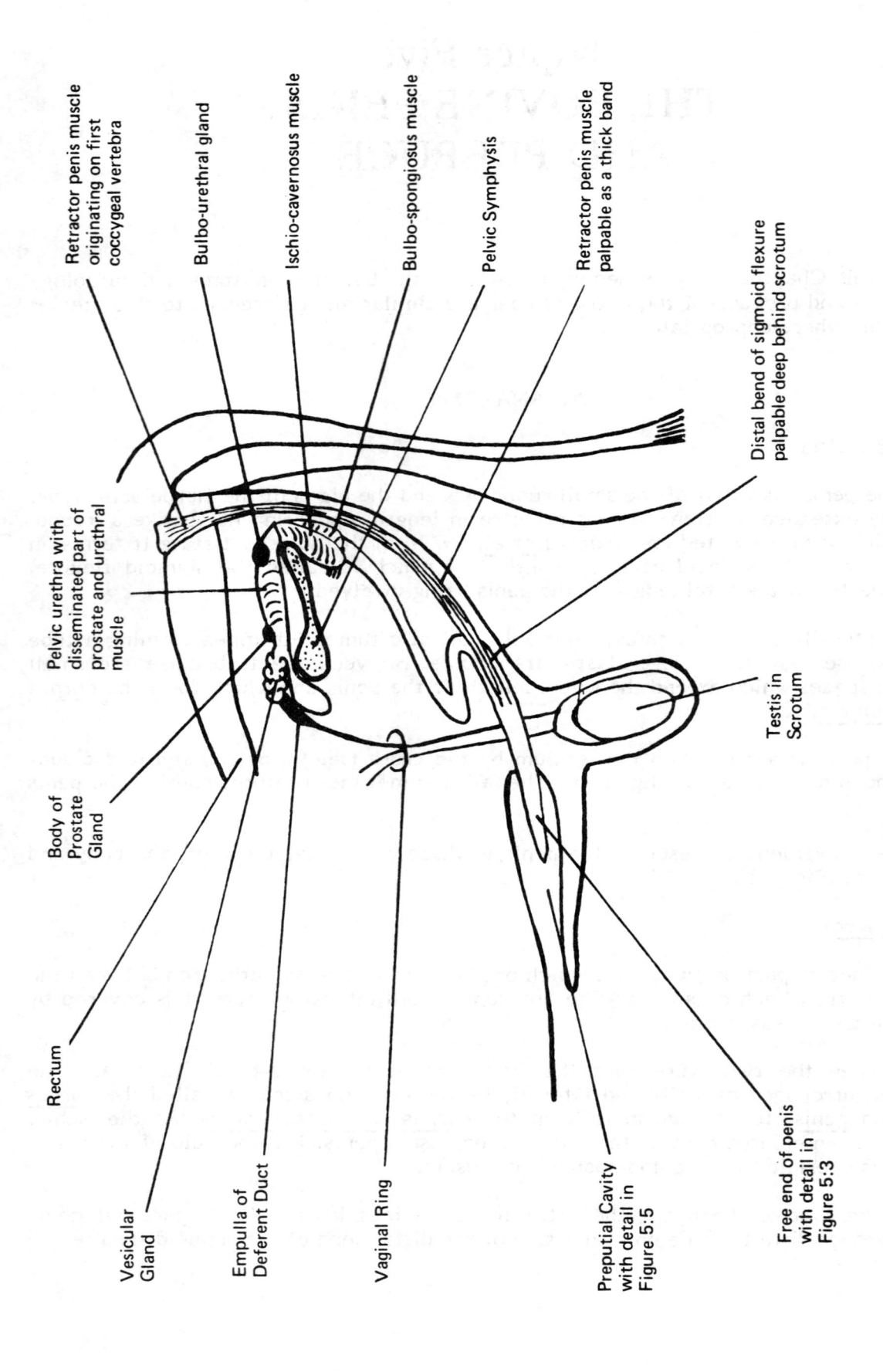

Fig. 5:1 Gross Anatomy of Bull Reproductive Organs.

(b) **The Body**

This portion of the penis is roughly oval in cross section (Fig. 5:2) and is formed by fusion of the crura with each other and with the distal continuation of the urethra and corpus spongiosum. Unlike the horse, the tunica albuginea of the ruminant and pig surrounds both the corpus cavernosum and corpus spongiosum, although these two structures do not become confluent.

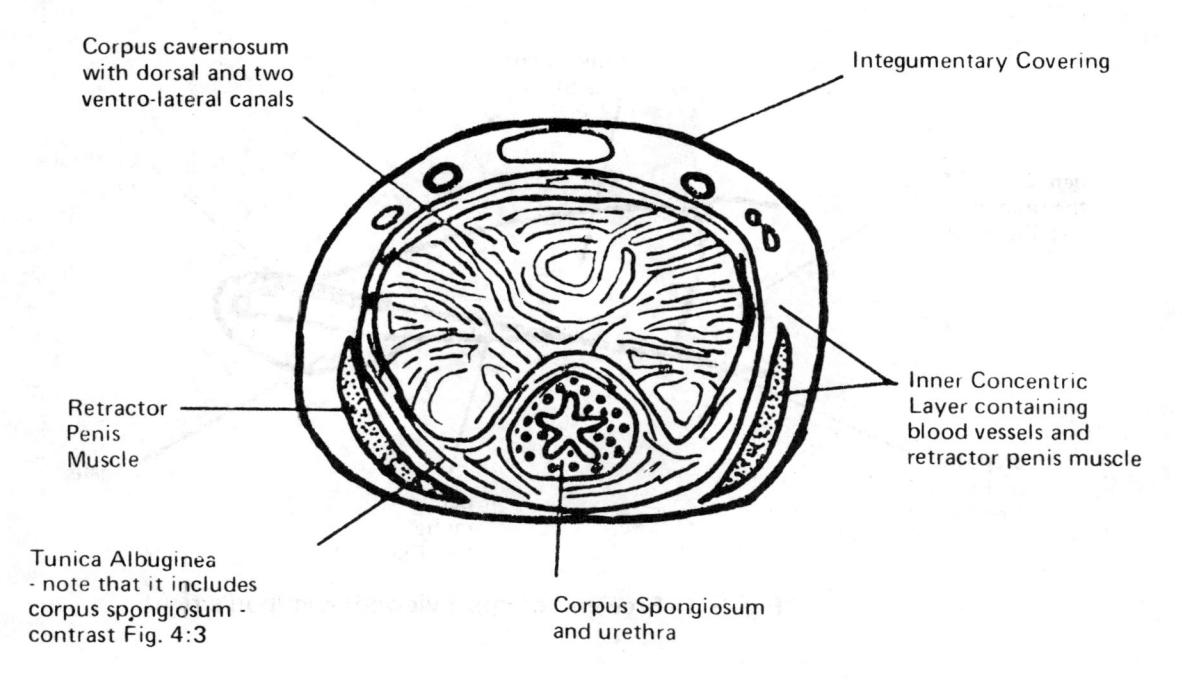

Fig. 5:2 Cross section through bovine penis about level of distal sigmoid flexure with penis fully extended.

In the body of the penis, the trabeculae of the corpus cavernosus penis are arranged about a single dorsal canal which terminates at the level of the distal bend of the sigmoid flexure, and two ventral canals, which can be traced to the distal extremity of the organ. These canals communicate with the cavernous spaces which, at the distal bend of the sigmoid flexure, are larger than normal and so at this point, allow both the straightening out of the sigmoid flexure at erection, and also encourage dorsal collapse during detumescence (Ashdown, 1970). Blood can flow either distally or proximally in these cavernous spaces. Dorsal to the body and <u>outside</u> the tunica albuginea lie some large dorsal veins which drain into the external pudendal vein.

(c) <u>The Free Part of the Penis</u> (Fig. 5:3)

This is the part of the penis which in the adult male lies within the prepuce. Generally, it is a continuation of the body. In the bull the apex is capped by a cushion of softer tissue, the asymmetrical and spirally twisted glans penis. A thin layer of erectile tissue (continuous with the corpus spongiosum) invests the glans penis - this is the corpus spongiosum glandis.

The free end of the penis is the predilection site for genital fibropapillomas in the bull (see page 109) and in some cases these tumours may be so large as to cover most of the free end. During surgery on these papillomas, the urethra should be identified. The urethra emerges from the corpus spongiosum about 4 cm from the end of the penis and runs in a groove on the <u>right</u> latero-ventral surface. The urethral orifice is very small - it will only admit a small to medium probe.

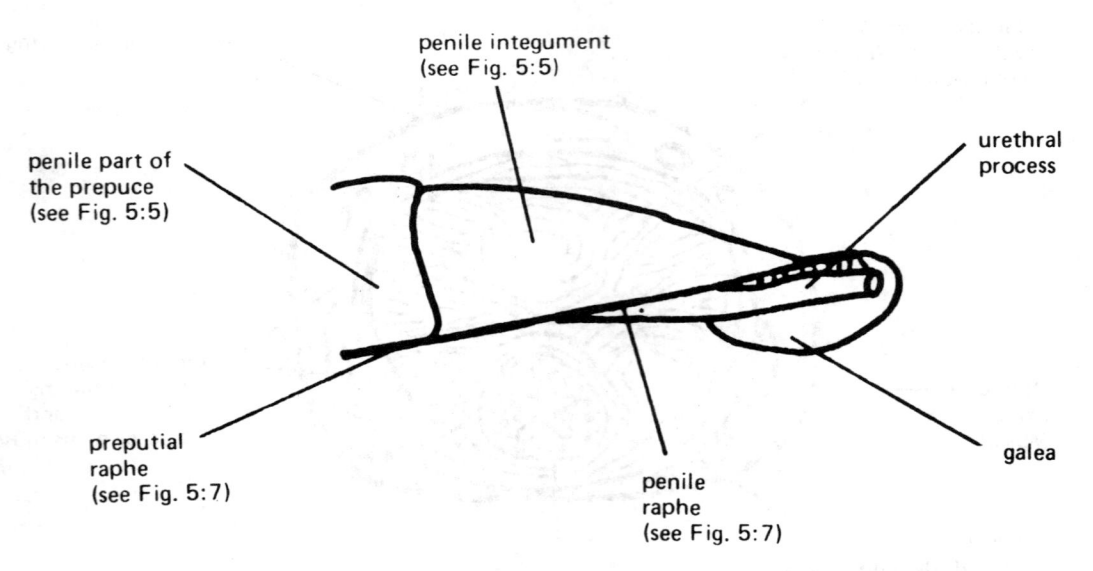

Fig. 5:3 Free end of penis viewed from bull's right.

The penile raphe is clearly seen on the ventral surface of the free end of the penis. It is the remains of the frenulum which was present in early post-natal life. (see page 98).

In the <u>ram</u>, the urethral process projects, worm-like, 3-4 cm beyond the end of the penis, becoming progressively narrower towards its extremity. The galea glandis is relatively larger than in the bull and has the appearance of a lizard's head. Immediately behind the galea lies a second swelling, the tuberculum.

Figure 5:4 shows the anatomy of the dorsal apical ligament, sometimes simply called the apical ligament. This structure originates from the tunica albuginea some 20 cm from the tip and inserts onto the tip of the tunica albuginea. It is separated from the underlying penile tunic by loose connective tissue except at its origins and insertions. If, therefore, the penile tip deviates in any way, this ligament is free to move and this has important consequences in the development of spiral deviation - see pages 100ff.

II   THE BLOOD SUPPLY AND ERECTION

The corpus cavernosum penis is supplied through its crura with blood from the deep artery of the penis which is a branch of the internal pudendal artery. (Occasionally in the bull, but frequently in the goat, the external pudendal artery appears to contribute to the deep artery of the penis). There are no direct venous connections between the corpus cavernosum and the dorsal veins (Ashdown and Gilanpour, 1974). Drainage of blood from the corpus cavernosum penis is, therefore, entirely through the deep veins of the crura to the internal pudendal vein. This venous drainage can be occluded by contraction of the ischiocavernosus muscles.

In contrast, the corpus spongiosum is drained at its distal end into the conspicuous dorsal veins. There are no vascular connections between the corpus cavernosum and the corpus spongiosum.

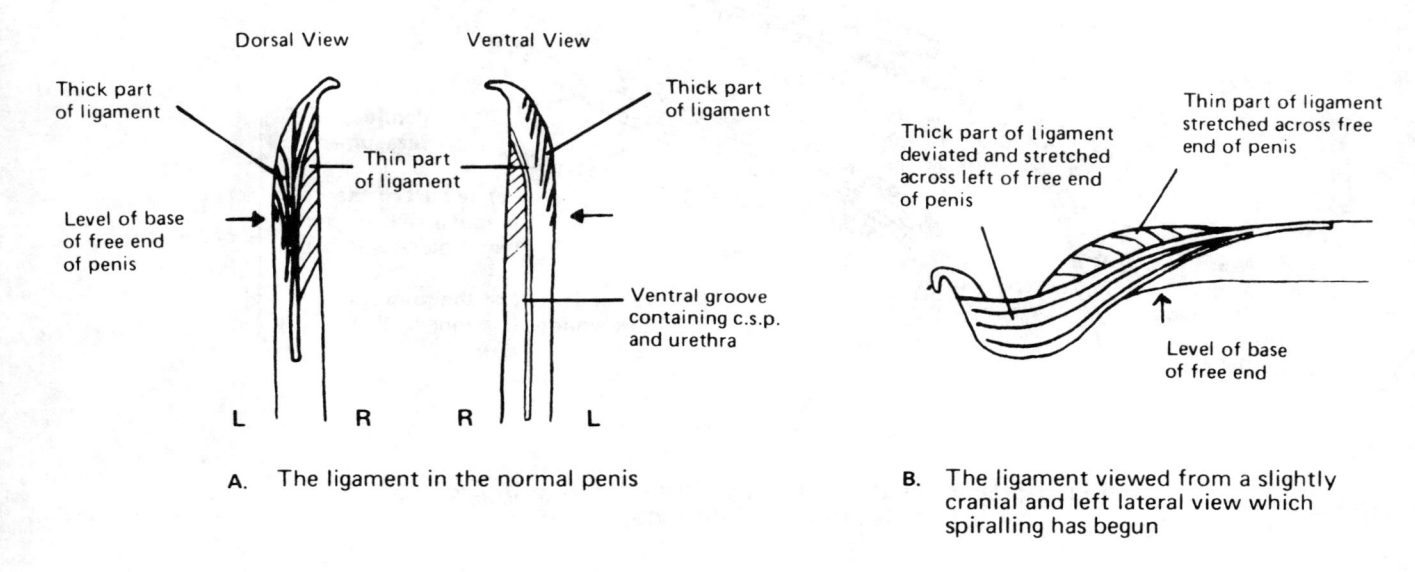

A.   The ligament in the normal penis

B.   The ligament viewed from a slightly cranial and left lateral view which spiralling has begun

Fig. 5:4   The dorsal apical ligament and its role in spiral deviation. The integument and hypodermis have been removed. (After Ashdown and Smith, 1969)

Erection is, therefore, initially due to an increased blood supply to the corpus cavernosum penis as a result of arterial vasodilation. When the bull is teased, the pressure inside the corpus cavernosum rises to about 100 mm of mercury (approximately systolic carotid artery pressure) and the penis moves forward so that the tip lies at about the level of the preputial opening. Full erection is the result of rhythmical contractions of the ischiocavernosus muscle which completely occlude drainage through the deep veins of the crura and which also pump blood from the crura into the distal closed system of the cavernous spaces of the body and free end of the penis. Pressures of up to 100 x arterial pressure are achieved at the point of maximum thrust (Beckett, Walker, Hudson, Reynolds, and Vachou, 1974).

Detumescence is probably due to cessation of activity in ischio-cavernosus muscle so reducing pressure in the corpus cavernosum and allowing blood to drain away normally. The sigmoid flexure re-forms because of the architecture of the fibrous structures (tunica albuginea and its trabeculae). The retractor penis muscle may play a minor role in returning the penis to the prepuce, but is not solely responsible.

The significance of this anatomical arrangement is discussed below in relation to impotence (page 114).

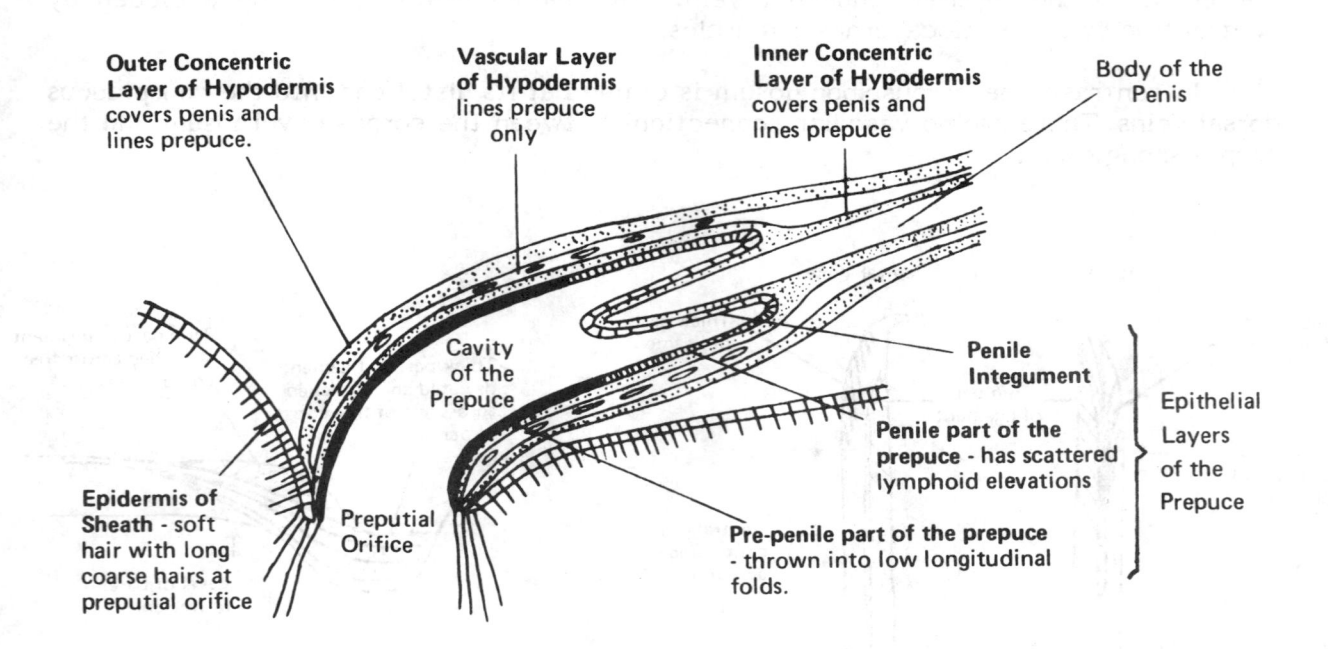

Fig. 5:5   Fascial and Integumentary Coverings of the Penis.
(After Ashdown and Pearson, 1973)

III   THE PREPUCE AND SHEATH

Technically, the term 'prepuce' includes both the tissues lining the preputial cavity and the fold of hairy skin outside (Nomina Anatomica Veterinaria). Whilst this concept might be useful anatomically, it is confusing clinically since various conditions affect the lining of the preputial cavity but not the hairy skin. Throughout this chapter, therefore, the term 'prepuce' refers to the hairless epithelium lining the preputial cavity and the term 'sheath' refers to the fold of hairy skin in which the prepuce lies. The two are continuous at the preputial orifice.

Fig. 5:5 shows the fascial layers and integumentary coverings of the penis and prepuce and gives the nomenclature used throughout these notes (adapted from Ashdown, Ricketts and Wardley, 1968, and Ashdown and Pearson, 1973a).

During early post-natal life, the penile integument and penile part of the prepuce are fused - see below under Persistent Penile Frenulum (page 98).

During erection the inner and outer layers of the preputial hypodermis slide across each other. The outer layer hardly moves at all but the inner layers slide out through the preputial orifice with the everted prepenile part of the prepuce. This sliding action is significant in considering the treatment of penile haematoma - see page 105.

Fig. 5:6 shows the preputial muscles which used to be called protractors and rectractors of the prepuce.

The cranial muscle elevates the pendulous part of the sheath and constricts the preputial orifice. The caudal muscle pulls the outer layer of the prepuce in a caudal direction but has no effect on the inner layer or on the penis. It is absent or reduced in polled Devons and Herefords and in the Angus bull. Otherwise it is readily palpable (5cm wide, 2-3cm thick) especially in young bulls.

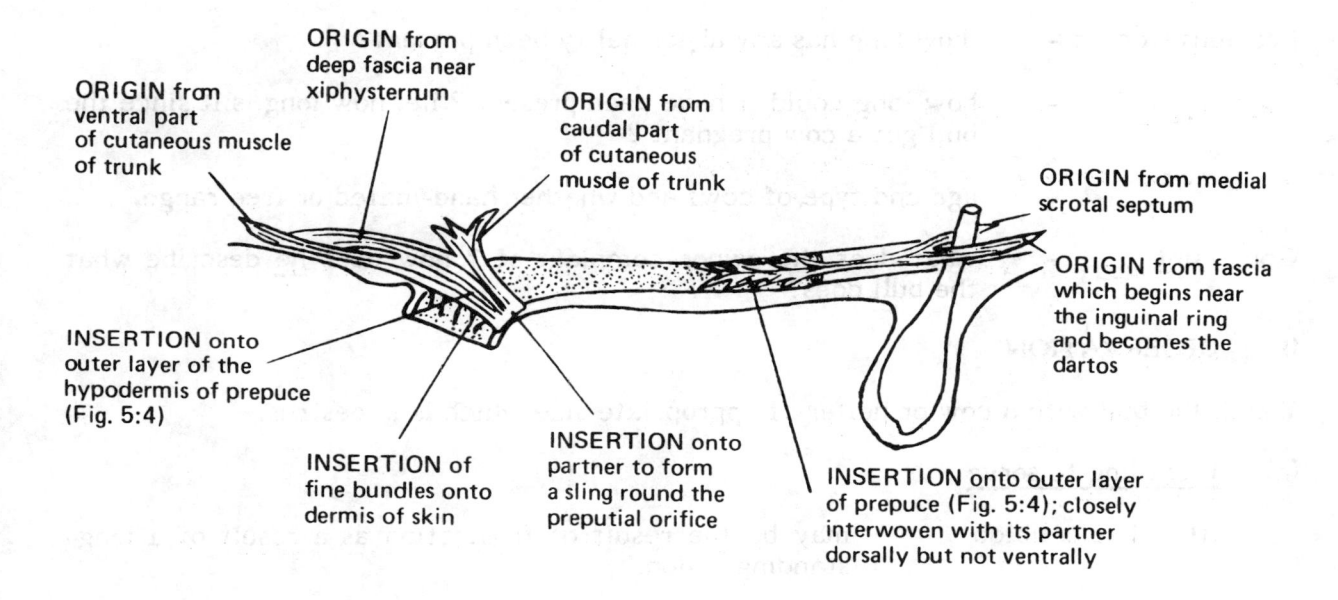

Fig. 5:6 Caudal and Cranial Muscles of the Prepuce.
(After Ashdown and Pearson, 1973a)

Both these muscles are supplied with motor innervation from the lateral thoracic nerve which originates from the median and ulnar cords of the brachial plexus. In some bulls the preputial orifice is close to the body wall, is directed cranio-ventrally and is tightly sealed. In other bulls, the pendulous part of the sheath is long and the orifice points ventrally and is always open. The significance of these muscles and the pendulousness of the sheath is discussed below under Prolapse of the Prepuce (page 111).

B. INVESTIGATION

This section is not concerned with investigations of semen quality in the bull but with conditions which either arise in the penis or prepuce, or may result in failure to achieve intromission. (Inability to serve a cow is not impotence - impotence strictly means failure to achieve an erection - see page 114)

I HISTORY

Age of Bull — may help to separate, congenital from acquired lesions but some 'congenital' lesions do not manifest themselves in young bulls, e.g. impotence due to vascular leakage (page 114) or spiralling (page 100).

Previous Record — how long has any abnormality been present?

- how long could it have been present? i.e. how long is it since the bull got a cow pregnant?

- age and type of cows and whether hand-mated or free range.

Complaint — what does the owner complain of? how does he describe what the bull does?

II OBSERVATION

Watch the bull with a cow or heifer of appropriate size which is in oestrus.

(a) <u>Unwilling to serve</u>

    (i) Lacks libido — may be the result of frustration as a result of a long-standing lesion.

                        — may be the management that is at fault, e.g. poor feeding, over feeding, over work.

    (ii) Has libido but is hesitant — usually the result of pain, possibly from a locomotor lesion, but some beef bulls may show delay or unwillingness in the presence of observers, but do not lack libido and are not in pain.

         Consider — reticulitis
                          rupture of lumbo-dorsal fascia;
                          arthritis;
                          penile tumour (pages 109-110)
                          hip dysplasia

(b) <u>Willing to serve</u>

(i) Penis fails to exteriorise or does so only slightly:

    Consider - Failure to erect (impotence) - see page 114..

- Penis or adnexa are too short: - see pages 97 and 110.

- Phimosis: - see page 100.

- Spiral deviation occurring within prepuce: - see pages 100-101.

- Penile haematoma which has resolved with peripenile adhesions: - see pages 103-106.

- Stenosis of preputial orifice: - see pages 111-113.

- Stenosis or adhesions proximal to preputial orifice: - see pages 107 and 111-113.

- Large tumour: - see pages 109-110.

(ii) Penis exteriorised completely but no intromission:

Consider -   Persistent penile frenulum: - see page 98.

- Spiral deviation which may be complete spiral or to right and ventrally only: - see pages 100-101.

- Ventral deviation either *per se* or following failure of operation to correct spiralling: -see pages 100-101.

- Service position not satisfactory (back remains almost parallel to ground) and/or searching movements lack purpose - may be idiopathic or due to locomotor lesion.

(iii) Penis exteriorised completely, intromission achieved but not thrusting:

Consider -   damage to dorsal nerve of penis (which is afferent pathway in thrust reflex) due to

(a) balanitis - especially following strangulation: - see page 106.

(b) haematoma which disrupted nerve - see pages 103-106.

(c) sequel to operation on dorsal apical ligament for correction of spiral deviation (rare).

(d) unknown causes.

III EXAMINATION

(a) General clinical examination first. A squeeze chute with a dropside is helpful. If a bar is placed behind the bull, the bull's ability to kick forward is reduced.

(b) Look for - discharge from preputial orifice = balanoposthitis (pages 102-103).

- preputial prolapse or eversion (pages 111-113).

- abnormal swellings within the sheath
  - abscessation
  - haematoma
  - tumours
  - ruptures of urethra

(c) Feel for - abnormal lumps on the body and free end of the penis, for size, presence of heat, painfulness, texture;

- the penis and test its mobility within sheath, especially to try to ascertain which layers are involved in any adhesion present; the penis of the conscious bull can be exteriorised by a combination of pushing back the sheath and pushing forward the penis; once exteriorised it may be held out by a surgical sponge.

- abnormal feel of stenosis in preputial orifice or preputial cavity (palpation may be facilitated by inflating the preputial cavity with air or oxygen, (Hofmeyr, 1967a)).

(d) Pudendal Nerve Block (Appendix III).

This is especially useful in cases where the penis does not naturally exteriorise completely. Under a successful pudendal nerve block the glans is always desensitized, so if it is possible to grasp the glans without the bull objecting, to apply traction, and still be unable to exteriorise the penis, then some physical impediment to penile extrusion is present. This could be stenosis or adhesions of the prepuce or peripenile tissues, myopathy of the retractor penis muscle or the congenital conditions discussed on page 97.

IV GENERAL CONSIDERATIONS

The bull often makes a very significant contribution to the gene pool of cattle. Some conditions of the penis and adnexa are congenital and may be hereditary, whilst other conditions certainly have pre-disposing factors which are heritable. Whether treatment of cases where genetic transmission may occur, is justifiable, therefore, is a matter of some debate.

Hofmeyr (1968a) has pointed out that no perfect cattle exist and, therefore, every fertile mating transmits some undesirable characteristic. Even when a condition has been shown to be heritable, he argues, it is still necessary to balance the disadvantage against the advantages. He points out that some conditions are stated to believed to be of genetic origin without submission of adequate factual evidence, a statement still true today.

Nevertheless, the veterinarian does have a responsibility to the cattle industry to protect it from undesirable features and, in particular, he has a responsibility to refrain from operations which may deceive a buyer of the bull's true heritable characteristics.

In conclusion, it is doubtful whether any operation to correct a congenital anomaly can be justified in a bull destined for the A.I. service. It is preferable not to operate on bulls being used in a bull breeding programme and one should never operate on a bull whose father suffered from the same complaint. In the case of bulls whose total progeny will go for beef operations it can be justified, but, ironically, such bulls are likely to be of little value.

C. SPECIFIC CONDITIONS

I CONGENITAL ABNORMALITIES

(a) <u>Congenitally Short Penis</u> ('Infantile Penis')

In the penis of the new-born bull the sigmoid flexure is poorly developed, the curves beginning to form at about 3 months of age. Development accelerates at puberty but full maturity is not reached until two years of age.

In some bulls, however, the penis does not grow fully and a variety of syndromes result. Some bulls may be able to serve as youngsters but as they mature, their abdomen becomes deeper, they become less agile and eventually copulation becomes impossible as the penis cannot reach the vulva of the cow. Other bulls may not be able to serve at all even as youngsters. In extreme cases no sigmoid flexure is present, whilst in others under pudendal nerve block, traction fails to exteriorise the penis even though the flexure has been eliminated.

These cases must be distinguished from those with true impotence (failure of erection).

No young bull unable to copulate should be rejected until he is two years old unless true impotence is proven or careful examination under pudendal nerve block suggests an extreme disability.

Some authors consider the disorder has an hereditary component (Roberts, 1971).

(b) <u>Congenitally short retractor penis muscle</u>

This condition results in signs similar to those exhibited in bulls with congenitally short penis. It has an hereditary background and is transmitted by a recessive character in Friesians (De Groot <u>et al</u>, 1946). Myectomy of the retractor penis muscle in the ischial area effected a cure in some animals but, because of the hereditary nature of the defect, such surgery is to be discouraged and, in fact, has been prohibited in Holland. The condition has sometimes been attributed to a fibrous metaplasia of the muscle.

(c) <u>Congenitally tight penile adnexa</u>

This condition has been described by Hofmeyr (1967b). The symptoms are similar to those of (a) above but, after passive controlled stretching of the penis under pudendal nerve block, normal extrusion and service was possible.

(d) Persistence of the penile frenulum

  (i) Development

The free end of the penis develops in embryonic and early post-natal life with its integument connected throughout its length, to the penile part of the prepuce by the ectodermal lamella (Ashdown, 1962, Fig. 5:7). At about 2 months of age the ectodermal lamella begins to split into two, forming the epithelia of the penile part of the prepuce and of the penile integument. Completion of this process of separation appears to be hormone dependant and may not occur in bulls until 8-11 months of age. Ventrally the ectodermal lamella is incomplete and here the bridge of connective tissue (= frenulum) must rupture. Excessive thickness of the frenulum, perhaps associated with the presence of a blood vessel traversing the frenulum, may cause persistence of the frenulum itself.

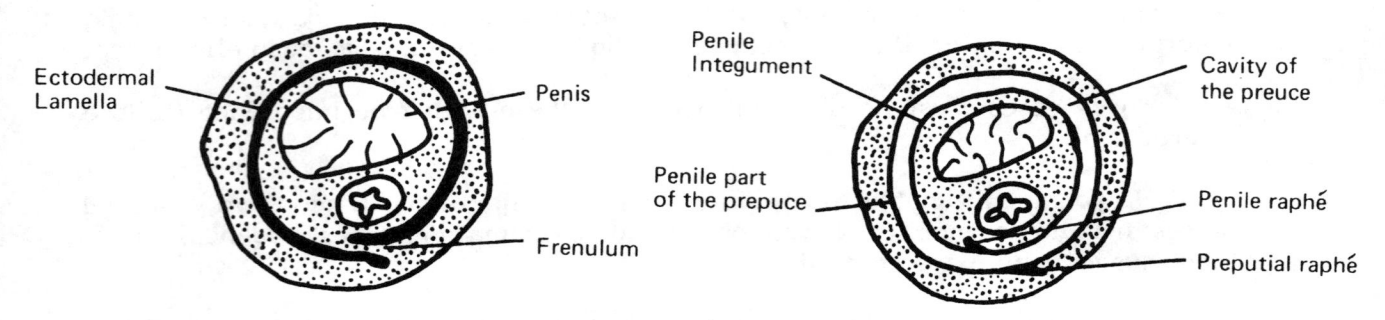

A. Cross-section of developing penis prior to splitting of the ectodermal lamella

B. Cross-section of mature penis after splitting of the ectodermal lamella (compare with Fig. 5:5)

Fig. 5:7 The penile frenulum. (After Ashdown, 1962)

  (ii) Clinical Signs

If the frenulum has persisted, a band of fibrous tissue extends from near the tip of the penis to near the junction of the penile part of the prepuce and pre-penile sheath. This results in a sharp ventral deviation of the erect free end of the penis (Fig. 5:8) and so prevents intromission.

(iii) Treatment

Carroll, Aanes and Ball (1964) found 40 instances in 10,940 bulls and showed it was most common in the Aberdeen Angus and Beef Shorthorn breeds. It has been known to occur in several of the offspring of treated Aberdeen Angus bulls (Copland, personal communication). It is, therefore, questionable whether treatment should be attempted. Treatment is, however, simple and effective. The bull is restricted in a chute, the penis exteriorised and anaesthetised. Ligatures are placed round the blood vessels and the fibrous structure is incised. The bull may be returned to stud work in 2 weeks.

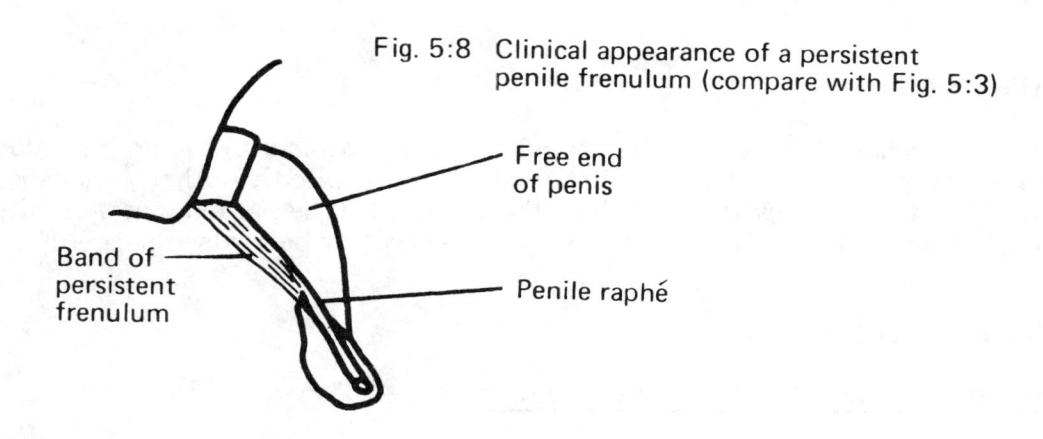

Fig. 5:8 Clinical appearance of a persistent penile frenulum (compare with Fig. 5:3)

- Free end of penis
- Band of persistent frenulum
- Penile raphé

(e) Other Conditions

Diaphallus or duplication of the penis is rare. In one reported case one penis contained a patent urethra and removal of the other may have effected a cure (Roberts, 1971). However, in other cases there is complete duplication of the urethra and treatment seems unlikely to be successful.

Hypospadias and Epispadias are congenital anomalies of the penis and urethra in which the urethra opens in the perineum, or on the ventral or dorsal surface respectively of the penis.

Intersex animals usually have ambiguous external genitalia and are easily recognisable clinically. Most are identified prior to adulthood and do not constitute a breeding problem.

II PHIMOSIS

The word 'phimosis' properly indicates a stricture of the preputial opening so that the penis cannot be extruded. Common usage, however, has extended its meaning so that it is often applied to conditions which lead to inability to protrude the penis.

Stenosis of the preputial orifice in bulls is usually acquired and is due to injuries, infection or surgery, often as a sequel to chronic preputial prolapse (see page 111).

Treatment of a stricture of the orifice involves removal of a wedge of preputial lining, sheath skin and intervening fascia on the ventral aspect just behind the preputial orifice. Lining and skin are then sutured together. Hofmeyr (1968b) reports that this procedure is often followed by the development of slight eczema at the point of the V as urine does not run away freely, and that although the opening is larger, some impedence to extrusion is still present.

III PARAPHIMOSIS

This is the term used when there is inability to withdraw the penis into the prepuce due to constriction of the preputial orifice but it has also been applied to other conditions which cause inability to withdraw the penis into the prepuce. True paraphimosis usually results in strangulation of the penis which is discussed separately below (see page 106).

IV DEVIATIONS OF THE PENIS

(a) Spiral Deviation (Also called 'Corkscrew' Deviation)

   (i) General Comments

   This type of penile deviation occurs in all breeds of bull at all ages. Although some cases have been seen in sons or closely related descendants of affected bulls (Roberts, 1971) the genetical basis of the condition is not yet clear.

   (ii) Aetiology and Clinical Signs

   The anatomical basis for the condition has been described by Ashdown and others (1967, 1968, 1969). They showed that in the penes of normal bulls the fibrous architecture of the tunica albuginea predisposed the penis itself to deviate ventrally and to the right when any external force was applied to the partially erect organ. Thus, when the body of the penis was pushed into its integument, the free end was initially deviated to the right (seen from behind) and ventrally. If further pressure was applied then the free end went into an anticlockwise spiral. During this process the dorsal apical ligament initially slipped to the left and ventrally; when the spiral was complete, the ligament lay across the spiral (see Fig. 5:4).

   Both the initial deviation and complete spiralling occur naturally in many normal bulls (Ashdown and others, 1968; Seidel and Foote, 1969). Its occurrence after intromission in normal bulls increases the area of contract between penis and vagina and may, therefore, increase tacticle stimulus and so promote ejaculation.

In affected bulls, however, deviation of the free end or complete spiralling occurs prior to intromission and so prevents coitus. It is possible that in these bulls, the penile integument is fully stretched over the penis (and so produces spiralling) too early in the process of copulation. Ashdown and Pearson (1973b) concluded that several different forces may operate to cause spiralling prior to intromission - affected bulls may simply be those whose penes spiral earlier than normal bulls. Others (e.g. Walker, 1970) have postulated a deficiency in the dorsal apical ligament; however, Ashdown and Pearson (1971) have shown that although removal of the ligament causes deviation of the free end ventrally and to the right, spiralling before intromission did not follow.

The condition presents itself in a number of ways, although in all cases the direction of the deviation and the spiral is the same. In some bulls the penis may spiral within the prepuce, sometimes preventing protrusion, whilst in others the spiralling may only occur when the cow's hindquarters are touched. Some bulls show deviation at almost every erection whilst others do so rarely and thus maintain at least a low level of fertility in the field. It is, therefore, important to observe the bull's mating behaviour on several occasions (see Blockey and Taylor, 1984). The onset of the condition may be gradual or sudden, may be associated with a change in management, or may appear in young bulls which have never served normally or in older bulls after several seasons of normal use. The condition is probably more common than is generally realised.

(iii) Treatment

Ashdown and Coombs (1968) showed that suturing the dorsal apical ligament to the tunic of the penis markedly reduced the ability of the penis to spiral in their experimental preparations. These findings confirmed the observation of surgeons that anchoring the dorsal apical ligament to the tunic (Walker, 1964) or strengthening it by fascial transplant (Hofmeyr, 1968c) often prevented spiralling in clinical cases and allowed normal intromission. The bull should be witheld from breeding for at least 2 months.

General anaesthesia is indicated. The penis is exteriorised, and held out by a towel clip and prepared for surgery. A 20cm incision is made on the dorsal surface of the penis (the urethra lies ventrally pushing from left to right) starting about 2cm from the tip. The incision is extended through the fascial and elastic layers down to the apical ligament, which is then incised in its thickest part for the length of the incision, thus exposing the tunica albuginea. Two large veins lie to the right and ventrally and they must be avoided. If simple suturing is employed, alternative sutures of catgut (to provoke reaction) and stainless steel (for strength) should be used to anchor the dorsal apreal ligament to the underlying tunic (Pearson, personal communication). If the fascial implant, is employed, the implant (3 x 12cm previously harvested) is sutured into place between the apical ligament and the tunica and anchored to both (Wolfe, Hudson and Walker, 1983). In both cases, the integument is sutured over the top and the bull denied sexual excitement for 60 days.

(b) Other deviations

Deviation occurring in the body of the penis rather than the free end have also been described but are less common.

Arthur (1975) describes cases in old fat Jersey bulls in which the abdomen was so large that the penis was always directed too far ventrally. The only possible treatment seems to be to place the cow in a pit!

Several authors have described bowing of the penis at attempted service in a vertical (= 'Rainbow' deviation) or horizontal plane or even into an S-shape. These deviations are sometimes attributed to an earlier injury which has healed by fibrosis and which prevents full extension of the penis on that side (Beckett, Walker et al (1974) have shown that, contrary to popular belief, the bovine penis does increase in length as erection by about 15%). Others have attributed the conditions to deficiencies in the dorsal apical ligament due to stretching, laceration and deterioration. Attempts to treat these deviations reflect these conflicting ideas of the aetiology. Thus, one school aims to shorten the convex side of the bend by removing A or D-shaped pieces of penile integument (Hofmeyr, 1968c) and tunic (Milne, 1954). Others (e.g. Walker, 1970) treat the conditions by one of the procedures described for anchoring or strengthening the dorsal apical ligament. Success rate is low - probably 25% at best will return to normal service. Boyd and Henselka (1972) described a technique which involved implanting a prosthetic device between the dorsal apical ligament and the tunica albuginea. Although they claimed success in 18 out of 20 bulls, they gave no details of the follow up procedure or the time involved. Moreover, they claimed that in cases of ventral deviation, the dorsal apical ligament was inadequately adherent to the underlying tunica albuginea of the penis, but as noted above (page 90) this is the normal situation.

V BALANOPOSTHITIS

Balanitis is an inflammation of the free end of penis and posthitis is an inflammation of the prepuce but because of the close proximity of the penis and prepuce, both organs are affected simultaneously, i.e. balanoposthitis is present.

(a) Infectious Pustular Balanoposthitis (IBR-IPV)

This is a disease in which there is acute inflammation followed by necrosis and sloughing of the epithelium of the free end of the penis and the prepuce. It has been shown by Bouters and others (1960) that this disease is caused by the herpes virus of cattle - Infectious Bovine Rhinotracheitis (IBR) Infectious Pustular Vulvovaginitis (IPV). (Although these viruses can be shown to be different by electrophoresis, they are serologically identical. It is usual therefore, to talk of the IBR-IPV virus). The virus produces ulcers and these become infected by the bacterial flora of the prepuce, resulting in severe pustular balanophosthitis. Affected bulls may spread the disease to cows by copulation, or by artificial insemination, so producing pustular vulvovaginitis. Viruses can usually be recovered from affected bulls for 10-14 days, although in some cases the virus can be recovered from the preputial cavity of a bull for periods up to one year (Snowden, 1965).

Epizootics of the disease occur and there is a great danger of this virus being disseminated in serum from artificial breeding studs (Roberts, 1971). Straub (1970) prepared a highly attenuated vaccine that would protect cows and bulls if instilled intravaginally or into the prepuce. The local immunity produced by this route of administration is better than that produced when the vaccine was given intramuscularly but a virulent IBR-IPV vaccine is available in some countries, but not in the U.K. In any case, careful thought should be given before this vaccine is used, as the herpes viruses tend to persist in the body and recurrent dissemination of virus may occasionally be possible.

(b) Granular venereal disease

This is the term given to hypertrophy of the lymphoid nodules on the penis and penile part of the prepuce. It is quite common but is not associated with any specific organism.

Treatment usually consists of sexual rest and the intra-preputial infusions of bland oily antibacterial solutions.

(c) Non-specific balanoposthitis

A number of authors have described ulceration of the pre-penile part of the prepuce which did not conform to the classical picture of IBR-IPV virus infection (Long and Dubra, 1972a; Pearson, 1972), but the significance is not clear.

Infectious posthitis is a serious problem in rams and wethers in Australia (Southcroft, 1965) and has been reported once in the U.K. (Doherty, 1985).

Non-specific balanoposthitis due to trauma is discussed below (page 107).

VI TRAUMATIC LESIONS

(a) Penile Haematoma

   (i) General Comments

This condition, also called 'Ruptured Penis', Broken Penis', 'Fracture of the Penis' or 'Rupture of the Corpus Cavernosum Penis', is undoubtedly the most important traumatic condition of the bovine penis. The haematoma develops on the dorsal aspect of the distal bend of the sigmoid flexure following rupture of the tunica albuginea of the corpus cavernosum. This specific haematoma should be distinguished from the more general haematoma that are described below under (c).

   (ii) Aetiology and Pathogenesis

In its classical form the haematoma is acquired during service. The site of rupture had suggested to earlier workers that the condition was associated with a tearing of the area where the retractor penis muscle is inserted. However, experiments by Beckett, Reynolds and others (1974) have shown experimentally that rupture of the tunica albuginea in this area can be induced by pressure within the corpus cavernosum penis of 2 to 4 times the maximum recorded in normal copulation. Since relatively small changes in the volume of the corpus cavernosum penis produce marked changes in pressure, it seems likely that the rupture occurs in living bulls at this weak point when sudden angulation of the penis takes place. Such angulation may occur when a cow moves suddenly at the time of maximum thrust by the bull. The possibility that prior weakening of the fibrous tissue plays a contributory role cannot be discounted.

Young bulls seem to be more prone to this injury (and also to other penile injuries) perhaps because of their enthusiasm and inexperience.

(iii) <u>Clinical Signs and Diagnosis</u>

In the classical form the bull shows shortening of the stride and stiffness, with a noticeable, but usually temporary, oedematous eversion of the prepuce. A swelling develops immediately cranial to the scrotum. At first the swelling is soft and fluctuating. It may be tender but it is not usually hot though skin discolouration may be marked. Its maximum size is reached at 24 hours after occurrence, after which it slowly shrinks as serum is absorbed and it becomes firmer as organisation occurs. Urination is not affected.

The presence of heat in the early stages indicates that an abscess is developing. Early or late abscess formation is said to occur (Pattridge, 1953; Pearson, 1972) in about half the cases, probably by haematogenous spread. Affected bulls may show an initial reluctance to serve, but this wanes. The bull will later attempt service but will be unable to extrude his penis fully or at all. The bull may not always be presented in the early stages, and the swelling of a resolving haematoma should be differentiated from an abscess in the sheath, and from neoplasia.

(iv) <u>Prognosis</u>

The prognosis depends upon many factors.

(i) Damage to the dorsal nerve of the penis (either at the time of injury or as a result of fibrous tissue impingement) with consequent loss of sensation of the free end occurs in some cases.

(ii) Infection may become established in the haematoma or surrounding tissues by haematogenous spread or surgical sepsis. Abscessation and/or adhesions may develop.

(iii) Disruption of the telescoping planes of connective tissue which heal by fibrosis and so prevent extrusion may be important, but Vandeplassche and others (1963) observed adhesions in only a small proportion of simple cases which did not recover. They noted that some bulls with 'thickened' tissues were able to serve normally, whilst most bulls that did not recover simply seemed unable to achieve erection. Young and others (1977) found acquired vascular shunts between corpus cavernosum and dorsal veins in some bulls with a history of traumatic injury (see also p.114 on Impotence).

Hofmeyr (1967c) suggests that if the penis can be exteriorised completely and there is sensation in the glans, then the prognosis is good, whilst in all other circumstances the prognosis is poor. Walker (1970) states that if the penis erects and extends fully when the bull is electro-ejaculated, the case is one in which surgery is indicated (but see comments on surgery below).

(v) Treatment

This condition has been the subject of many different treatments. Although Vandeplassche and others (1963) suggested that 50-60% of cases recovered after 4-6 weeks of rest, Ogden (personal communication) suggests that the percentage is much lower (15-20). Current thinking is that surgical evacuation of the clot reduces the time taken for complete reorganisation, especially if the haematoma is large, lessens the risk of abscessation, and allows free penile movement before adhesions have formed. However, it should be accompanied by steps to prevent the development of adhesions between the sliding planes of fascia. Although teasing the bull daily post-operatively with a cow in oestrus will encourage penile movement, Hofmeyr (1967c) has pointed out that gentle massage for 10-20 min daily provides more movement than numerous teasings and that such movement is independent of the ability of the penis to respond to sexual stimuli.

(1) Conservative Treatment. Systematic antibiotics are widely advocated to control infection and the development of abscesses. However, it is likely that the suppurative organisms are already present in the clot and will not, therefore, be reached by systematic antibiotics. The application of ice packs or cold water hosing or of hot water several times daily has been held to be beneficial. After three days, daily massage of the affected areas is imperative - the skin over the area is grasped and moved firmly but gently back and forth, tensing without excessively straining the underlying tissues. 'No progress may be apparent for weeks, then suddenly, the penis can protrude spontaneously and normal erections may follow in one or two weeks' (Hofmeyr, 1967c).

(2) Minor Surgical Treatment. This treatment involves injecting proteolytic enzymes into the clot 5-7 days after formation (Metcalf, 1965). Antibiotic and 125,000 i.u. of a streptokinase/streptodornase preparation ('Varidase', Lederle Laboratories) are dissolved in 250ml of 0.9% saline and injected through the skin into about ten areas of the haematoma, care being taken to be as clean as possible about the procedure. Five days later, a suction tube is inserted through a stab-wound in the skin under conditions of surgical cleanliness and the now liquified haematomoa gently removed by suction. This treatment aims to make the smallest possible incisions across the telescoping fascial planes. Should abscessation occur Metcalf states that the abscess can be drained and the cavity irrigated daily through a similarly small wound for the same reason.

(3) Radical Surgical Treatment. This treatment, although widely advocated, has been severely criticised by Hofmeyr (1967c) on the grounds that it makes large incisions across telescoping layers of connective tissue and thus is more likely to result in the development of adhesions which restrict penile movement. Others suggest that surgical intervention early in the course of the condition gives the bull the best chance of recovery. The optimum time for surgery is about 1 week after the injury. Delayed surgery is rarely successful (Pearson, 1972). The bull is given a general anaesthetic or placed in lateral recumbency with heavy sedation and local infiltration anaesthesia.

Under surgical cleanliness of a high standard the skin is incised and the blood clot is exposed and gently removed, taking care to control all bleeding. Any adhesions that have developed should be incised and all tissue debris removed. The tear in the tunica albuginea should be sutured if possible although the corpus cavernosum should not be penetrated to reduce the risk of development of vascular shunts (see p.114). The subcutaneous tissue is carefully closed and the skin sutured. The bull is teased or massaged daily but not allowed back into actual service until full protrusion and normal erection are known to occur.

(iv) Haematomoa proximal to the sigmoid flexure

Hudson (1971) has described finding a haematoma in the dorsal longitudinal canal of the penis of one impotent bull. Cembrowicz (quoted by Ashdown, Gilanpour and others, 1979) has described two further cases with haematoma proximal to the sigmoid flexure and Ashdown and his co-workers have themselves described two more. The most striking presenting symptom in all cases was impotence and this is discussed further below (p.114). It is not known whether the haematoma were of traumatic origin.

(c) Strangulation of the Penis

Strangulation can occur in a variety of circumstances. These include:

(i) following collection of semen when a rubber band on the artificial vagina falls off onto the bull's penis (for this reason rubber bands should never be used on an artificial vagina);

(ii) in cases of true paraphimosis;

(iii) in animals in which a rubber band has been maliciously placed around the organ;

(iv) in bulls in which the preputial hairs have been cut short, hairs can trap a retracting penis so producing paraphimosis and a mild degree of strangulation.

(v) in young bulls reared in homosexual groups, riding of other bulls can accumulate on the penis of aggressive bulls and form a firm band which strangulates the penis.

With penile strangulation the distal portion becomes necrotic and gangrenous fairly rapidly, so the prognosis is always guarded. Rubber bands and hair rings must, of course, be removed. In cases of paraphimosis the penis must be returned to the preputial cavity after being cleaned, and the preputial orifice temporarily (24-48 hours) closed with a bandage round the sheath, a purse string suture or small towel clips. It may sometimes be necessary to enlarge the preputial orifice. Frequent massage and applications of antiseptic ointment are indicated for 1-2 weeks to prevent adhesions. In fresh cases the prognosis is good, but even a few hours ischaemia may produce enough pathology to prevent the bull ever breeding again.

(d) Non-specific trauma

(i) Aetiology

These injuries may occur at service if a cow moves suddenly when the bull is searching or thrusting or they may develop if the bull straddles a gate or fence, or when he rides other cows or cattle. A recognised injury in bulls collected by A.V. is a transverse tear at the junction of the penile integument and the penile part of the prepuce. Sexually enthusiastic youngsters seem particularly prone to develop a non-specific balanoposthitis. This may be mild or severe, according to the nature of the initial injury and how many females have been served before pain overcomes libido or the bull and his cows are separated - repeated attempts at service by the bull may exacerbate damage. Rams can also develop a non-specific balanoposthitis in their first season.

(ii) Pathogenesis

Three general types of injuries can be distinguished although they can and do occur together:

(1) Haematoma may develop (i) from the dorsal vessels of the penis (see Fig. 5:2); (ii) through a rupture in the tunica albuginea other than at the weak point of the classical haematoma referred to above; (iii) from the vessels of the vascular layer of the hypodermis (see Fig. 5:5); (iv) within the corpus cavernosum itself. Noordsy and others (1970) say that 50% of cases of haematoma in beef bulls occur in the vascular tissues outside the penis. These haematomas may, like those in the classical form, develop into abscesses or, if the tunica has ruptured, result in vascular shunts between corpus cavernosum and dorsal veins. If they occur in the vascular layers outside the penis they may resolve with the deposition of fibrous tissue between the inner and outer concentric layers of the hypodermis - normal penile protrusion is then prevented.

(2) Laceration and/or contusion of the preputial membrane may occur leading to a non-specific balanoposthitis. The normal preputial cavity of the bull contains a wide variety of organisms including E.coli, streptococci, staphyloccocci, P. aeroginosa, C. pyogenes, Proteus spp, Actinomyces, Actinobacillus, mycoplasms, IBR-IPV virus and occasionally Mycobacterium. Should any part of the lining of the prepuce be severely traumatised, then these bacteria can gain entry to the deeper layers of the sheath and set up inflammatory changes. In the mild case, contusion is superficial and an unwillingness to serve may be evident. Frank haemorrhage from the damaged preputial lining may occur in more severe cases as may necrosis and suppuration with a marked preputial discharge. Abscesses may develop and these may become walled off, temporarily or permanently, or burst either into the preputial cavity or through the skin to the outside. Where traumatised mucous membranes of the preputial lining are left in apposition they may heal by fibrosis producing adhesions or stenosis, either of which may prevent penile extrusion partially or totally. IBR is said to be particularly adept at causing ulceration and adhesions in cold climates such as Canada.

(3) <u>Urinary Tract Involvement</u>. The bovine urethra is extremely well protected except for its terminal 4 cm (see Fig. 5:2 and p.90) so it only becomes damaged in very severe cases in which other lesions are present. A possible sequel to urethritis is ascending infection with possible seminal vesiculitis and ampullitis.

(iii) <u>Clinical Signs and Treatment</u>

Bulls may be presented at any stage of the pathological process with a simple or a complex clinical picture. It is only if the bull is exceptionally valuable or if the owners are prepared for a prolonged period of nursing care that treatment of the complicated case is worth attempting.

Mild cases require a brief period of rest from sexual encounters.

In the early stages it is the lacerations which require intensive care. They must <u>not</u> be sutured as the wound can never be thoroughly cleaned. Initial gentle scrubbing with surgical soap should be followed by daily irrigation accompanied by thorough massage to ensure dispersal. Hydrogen peroxide may be used until foaming no longer occurs (Walker and Vaughan, 1980). Installation of an oily antiseptic material (e.g. 2g tetracycline and 60ml scarlet oil in 500g of anhydrous lanolin) may also be used. Daily massage of the penis backwards and forwards in the sheath discourages the development of adhesions. The bull should not be allowed back into service until the penis can be extended without tearing the prepuce - this may be at least 4-6 weeks.

If a haematomoa can be localised then it may be removed surgically as in the classical form.

Many cases, however, may not be presented until an abscess has developed or adhesions have formed with or without stenosis and it is realised that the bull cannot fully protrude his penis. In such cases excision of the scar tissue offers the only hope of restoring normal function. The technique of circumcision described under penile prolapse is the technique of choice (p.113).

Developed abscesses carry a poor prognosis. They should preferably be excised via the internal surface of the prepuce with the penis extended. If they are drained (or burst during attempted excision) then it is difficult to prevent adhesions developing which interfere with penile protrusion - Walker and Vaughan (1980) suggest establishing adequate drainage, allowing healing to occur and finally attempting excision of the adhesion some weeks later.

VII TUMOURS OF THE PENIS

Tumours may cause phimosis or paraphimosis or may prevent normal intromission. The only significant tumour is fibropapilloma, others being rare.

(a) Fibropapillomas

(i) Aetiology

There is evidence that penile papillomas and cutaneous warts are caused by the same viral agent but occasionally cutaneous lesions may occur without penile warts and penile warts without skin lesions. The condition may occur in epizootics with both cows and bulls affected or it may be limited to the bull. The lesion forms a single or mutiple cauliflower-like growths on the free end of the penis, almost invariably in young bulls (1-3 years), most usually in bulls in their second year. Spontaneous regression may occur and may be associated with the completion of puberty (Formston, 1953). Recurrence can be due either to inadequate removal of all the tumours or to new tumours growing.

(ii) Clinical Signs

There is usually haemorrhage from the prepuce after service although on some occasions the farmer may first notice bleeding from the cow's vulva after service. Discomfort or pain at intromission may cause hesitancy or refusal to serve, or even paraphimosis or phimosis. The lesion is almost invariably palpable through the sheath. Pearson (1972) has recorded several cases, some in steers, where the growth was responsible for urethral obstruction and led to rupture and extensive cellulitis due to infiltration by urine of the peripenile tissues.

(iii) Treatment

Because of the tendency to spontaneous regression, it is not always possible to make a valid assessment of any given treatment. The administration of Lithium Antimony Thiomalate enjoyed a vogue. Vaccines have also been much employed, but, although there is some evidence that homologous vaccines may be better than heterologous ones, Olson et al, (1960) believe that the effectiveness of the vaccine in treatment is extremely limited.

Surgical removal of small tumours is readily performed under local anaesthesia after snaring the penis when the bull jumps a cow, but as there is some evidence that application of a snare or tourniquet too tightly can permanently damage the dorsal nerve of the penis, pudendal nerve block (Appendix III) or general anaesthesia are preferable. Removal of the tumours, which are usually attached by a relatively small area, is most easily effected by a pair of curved scissors. Adequate removal of all tumour tissue is clearly desirable but care should be taken to avoid unnecessary damage to the urethra by identifying the latter with a probe or catheter. Even so, the urethral wall is sometimes involved in the tumour and has to be removed. (Hofmeyr (1968a) recommends actual creation of a permanent fistula but Pearson (1972) recommends leaving the wound to form its own fistula.) Haemostasis can be achieved by ligation or by a light touch with diathermy. Where possible the epithelium should be sutured over to speed healing. Haemostasis with the larger tumours is a problem. Ligation of the bleeding vessel is usually extremely difficult and attempts to suture over the mucous membrane may result in more haemorrhage. Direct electro-cautery may also produce more haemorrhage. Application of diathermy through a pair of haemostats clamped on the bleeding point is, in the author's experience, more satisfactory. Fortunately, the haemorrhage is seldom dangerous and usually slows down and eventually stops.

The sheath should be liberally filled with an antiseptic ointment. The animal should not be used again until erection occurs without bleeding - usually about two weeks. They should at least be teased at this stage to detect early signs of regrowth (Pearson, 1977).

Occasionally, amputation may be necessary if the growths are extensive and, provided only the terminal portion of the free end is removed, the animals may serve normally (Milne, 1954; Pearson, 1972).

(b)   Bleeding Haemangioma ('Erection Haemorrhages')

Small haemorrhagic fistulae communicating with the corpus spongiosum may be present on the free end of the penis as lentil-sized, reddish-blue openings which bleed during erection. They may be difficult to see in a flaccid penis but a light tourniquet on the body will often produce bleeding. Small amounts of blood (or on occasions a spray of blood) can sometimes be seen at ejaculation. Ashdown and Majeed (1978) have presented evidence that this blood originates from leaks from the corpus spongiosum into the urethra. The condition may resolve if service is prevented for six weeks or if the lesion (assuming it is accessible) is treated by electro-coagulation.

VIII   DEFECTS OF THE RETRACTOR PENIS MUSCLE

(a)   Congenitally short retractor penis muscle

See page 97.

(b)   'Contracted' retractor penis muscle

This is a term confined to states in which the retractor muscle has been affected by secondary changes. Hofmeyr (1967b) has given an account of atrophy of the muscle, apparently due to disuse after a long period in which erection has been discouraged or was impossible, such as might occur in an animal with arthritis or a stenosed prepuce. Konig (1961) has described a chronic non-purulent myositis characterised by pearly swellings, but these must be distinguished from the 'beads' seen in many bulls which represent local areas where the tonus of the muscle is high.

Affected bulls may, therefore, have served normally for years but later become unable to extrude their penis.

Myotomy of the muscle is the only possible treatment. It may be carried out at the ischial arch under epidural anaesthesia (Hofmeyr, 1967b) or just cranial to the scrotum under general anaesthesia (Ashdown and Pearson, 1973a). An emasculator is the simplest means of achieving haemostasis.

However, it must remain a matter of doubt whether defects of the retractor penis are at all significant. The muscle may be considerably shortened experimentally without interfering permanently with protrusion.

(c) Calcification of the retractor penis muscle

McEntee (1969) has described finding calcification of the retractor penis in old bulls but, underlining the point made in the previous paragraph, even though the muscle may be almost completely calcified, normal service has been possible.

IX PROLAPSE OF THE PREPUCE

(a) General Comments

This can be a serious disease of certain types of cattle. Although bulls of many breeds can suffer from this malady, it is more common in the Bos indicus breeds and the polled beef breeds of Bos taurus (e.g. Aberdeen Angus and Hereford). It is rare in dairy cattle. It was the subject of considerable research in the 1970's.

(b) Aetiology

Bellenger (1971) believes that a longer total preputial length is important in Bos indicus but Long and Hignett (1970) have shown that there was no difference in preputial length between Bos taurus breeds which evert the prepuce and those which do not. In contrast, Long and Hignett (1970) found that the caudal preputial muscle was virtually absent in Bos taurus breeds which everted the prepuce, whilst Bellenger (1971) found it was present in Bos indicus breeds which did evert the prepuce. Later, Ashdown and Pearson (1973a) showed that the caudal preputial muscles, even when present, are apparently without any effect on the prepuce. They have no meaningful attachment to the preputial lining and can be sectioned without inducing preputial eversion.

(c) Significance

This is also a matter of some debate. In all breeds, preputial eversion may be no more than a bad habit and, in a given environment, without serious consequences for the bull. There is some evidence of an increased risk of contamination of semen samples in bulls which evert the prepuce and this could be important in an artificial insemination programme (Meredith, 1970).

However, the prolapsed prepuce is clearly more likely to be damaged than when retained and, when bulls with a predisposition to eversion live in environments where the likelihood of damage is high, preputial prolapse becomes a very significant disease. Thus, it is important in countries where the bull live in scrub bush rather than on green pasture and in areas which are heavily infected with ticks, particularly those with long mouth parts. As a result of trauma, inflammatory oedema develops, and gravity and the oedema impede both venous and lymphatic drainage. The extra weight of the oedema induces further protrusion. Drying of the surface results in cracking, which allows bacteria to enter and gangrene may follow. The oedema may eventually obstruct the entry of arterial blood and ischaemia and necrosis will supervene. If left untreated, resolution by fibrosis may occur, but in such animals there is usually severe stenosis of the preputial orifice and resulting inability to serve.

Affected animals may, therefore, be seen as (i) preputial everters without any pathology, as (ii) acute or chronic cases of preputial prolapse or (iii) healed cases in which stenosis of the preputial orifice is the most significant clinical finding.

(d) Treatment

(i) Preventive Surgical Treatment. It is a simple matter to reduce the pendulousness of the bovine sheath by placing a tuck in the skin on either side (Fig. 5:9). Local anaesthetic and absorbable suture material are used. Lagos and Fitzhugh (1970) showed that there was a correlation between the pendulousness of the sheath and the extent to which the prepuce everted and that the later factor was highly hereditable. Although Long and Dubra (1972b) argued that frequency and duration of eversion were more important and showed that these were not highly hereditable, it is still doubtful whether any operation should be performed on animals whose sons will later be used in breeding programmes.

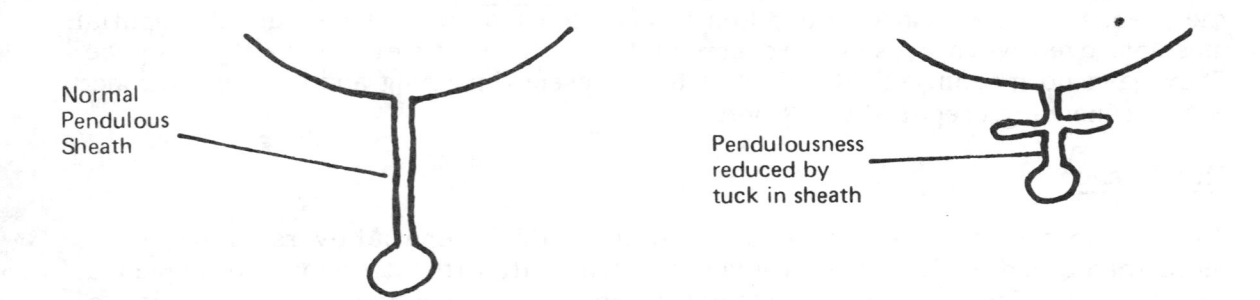

Fig. 5:9 Showing a simple tuck on each side of a pendulous sheath reducing the pendulousness.

(ii) **Conservative Medical Treatment.** A latex tube with a wall which is thick enough not to collapse under pressure, is inserted into the preputial cavity so that urine flow is not impeded. Exposed preputial epithelium is covered by a stockinette bandage. The prolapse and immediately adjacent sheath are snugly wrapped with elastic bandage. This pressure bandage is changed every two or three days until the oedema subsides. This type of treatment is applicable to early cases in which only a small proportion of preputial mucosa is permanently prolapsed. Early acute cases of simple prolapse respond rapidly to this sort of treatment. In more severe cases, conservative treatment is useful in reducing the swelling prior to undertaking corrective surgery (Larsen and Bellenger, 1971).

(iii) **Corrective Surgical Treatment.** The aim of circumcision is to resect entirely an annular area of affected integument and to join healthy preputial integument to healthy preputial integument. The technique described by Wolfe, Hudson and Walker (1983) aims to retain as much healthy tissue as possible and is relatively simple to understand, although its execution calls for patience as well as skill.

The bull is placed under a general anaesthetic and the penis is exteriorised and prepared meticulously for surgery. The total length of healthy pre-penile and penile parts of the prepuce must be at least twice that of the penile integument if the bull is to retain breeding potential. As it is essential that the preputial lining is not twisted, two marker sutures of suitable material are placed outside the area to be excised.

Circumferential incisions are made a sufficient distance apart to excise the affected tissue. The circumferential incisions are connected by a longitudinal incision and the affected portion of prepuce is dissected away, preserving as much elastic tissue as possible, but removing all diseased tissue whether abscessated or fibrosed. It is essential that careful attention is paid to haemostasis and all major blood vessels encountered during dissection should be carefully cauterised or ligated. It is most important that as much preputial lining as possible is preserved, but it is possible to resect skin from the preputial orifice.

Finally, the sub-integumentary layers are apposed with interrupted absorbable sutures. The preputial epithelium apposed also, employing the marker sutures to ensure that the preputial lining is not twisted.

A Penrose drain is sutured over the glans and is left long enough to come out through the preputial opening - this avoids urine contamination of the surgical field. A smooth ended perspex tube some 2" in diameter is inserted into the preputial orifice over the Penrose drain - the object is to prevent stenosis. An elastoplast bandage is applied over the sheath and exposed tube to hold the tube in position and to control oedema - the bandage may be tacked to the sheath proximally, but it and the perspex tube are removed after five days. Post operatively, a course of antibiotics is essential and the judicious use of corticosteroids appears to reduce post-operative swelling and the development of fibrosis without unduly delaying healing. The bull is sexually rested for 60 days.

X   IMPOTENCE

True impotence occurrs when penile erection is insufficient for intromission - the term impotence should not be used to describe other conditions. The bull's penis is too flaccid for intromission unless high pressures are generated within the corpus cavernosum (see page 91). These pressures may not be generated if there are possible escape routes for blood from the corpus cavernosum penis other than those at the crura. These may be shunts from corpus cavernosum to corpus spongiosum or to dorsal veins. Alternatively, a blockage in the proximal part of the penis may prevent blood reaching the distal part. If penile erection is insufficient for intromission then the bull is said to be impotent.

Recent anatomical, clinical and radiographic studies have considerably enhanced our knowledge of this condition in the bull (Young and others, 1977; Ashdown, David and Gibbs, 1979; Ashdown and Gilanpour, 1974; Ashdown, Gilanpour, David and Gibbs, 1979). True impotence also occurs in the boar (Ashdown, Barnett and Ardalani, (1982). Four categories of lesions may be recognised:-

(i)   The corpus cavernosum is drained, within the body of the penis, by a regular series of large distal veins, presumed to be congenital. Bulls with such lesions have never been able to serve.

(ii) The corpus cavernosum is drained by a distal network of small veins. These bulls may be able to serve as youngsters but become progressively unable to develop a full erection.

(iii) The corpus cavernosum is drained by veins that have developed at the point of traumatic injury to the tunica albuginea. Such bulls as these have a history of a traumatic incident such as penile haematoma (p.104) from which they appear on clinical examination to have recovered yet they are now impotent.

(iv) The dorsal (and possibly other) canals of the corpus cavernosum are occluded either by fibrous tissue (possibly of traumatic origin), haematoma or a thrombus (see Hudson, 1971). The possibility of turbulence causing mural thrombi exists. Other lesions may also be present in this type of impotence. See also page 106.

Impotent bulls may be presented, therefore, as cases which have never been able to achieve intromission (= primary impotence) or as cases which have become impotent after a period of successful service (= secondary impotence).

Diagnosis depends primarily on being able to show that, although the ischiocavernosus muscles are contracting (they can be palpated in the perineal region immediately below the pelvic brim as the bull mounts), the penis remains flaccid. Precise localisation of the lesion depends upon being able to take adequate radiographs in the live bull. (See Young and others, 1977 and Glossop and Ashdown, 1986).

Treatment of bulls in categories (ii) and (iv) is unlikely to be successful. Young and others (1977) have had some success in bulls with identifiable shunts by excising the vascular shunt, and suturing over the defects in the tunica albuginea.

D. REFERENCES

Arthur, G.H. (1975) Veterinary Reproductions and Obstetrics. Fourth Edition. Bailliere and Tindall, London.

Ashdown, R.R. (1962) Persistence of penile frenulum in young bulls. Veterinary Record 74 : 1464-1468.

Ashdown, R.R. (1970) Angioarchitecture of the sigmoid flexure of the bovine corpus cavernosum penis and its significance in erection. Journal of Anatomy 106 : 403-404.

Ashdown, R.R. (1976) The shape of the free end of the bovine penis during erection and protusion. Veterinary Record 99 : 354-356.

Ashdown, R.R., Barnett, S.W. and Ardalani, G. (1988) Impotence in the boar. 2: Clinical and anatomical studies on impotent boars.

Ashdown, R.R. and Coombs, M.A. (1967) Spiral deviation of the bovine penis. Veterinary Record 81 : 126-129.

Ashdown, R.R., David, J.S.E. and Gibbs, C. (1979) Impotence in the bull : (1) Abnormal venous drainage of the corpus cavernosum penis. Veterinary Record 104 : 423-428.

Ashdown, R.R. and Gilanpour, H. (1974) Venous drainage of the corpus cavernosum penis in impotent and normal bulls. Journal of Anatomy 117 : 159-170.

Ashdown, R.R. Gilanpour, H., David, J.S.E. and Gibbs, C. (1979) Impotence in the bull : (2) occlusion of the longitudinal canals of the corpus cavernosum penis. Veterinary Record 104 598-603.

Ashdown, R.R. and Majeed, Z.Z. (1978). Haemorrhage from the bovine penis during erection and ejaculation: a possible explanation of some cases. Veterinary Record 103: 12-13.

Ashdown, R.R. and Smith, J.A. (1969) The anatomy of the corpus cavernosum penis of the bull and its relationship to spiral deviation of the penis. Journal of Anatomy 104 : 153-159.

Ashdown, R.R. and Pearson, H. (1971) The functional significance of the dorsal apical ligament of the bovine penis. Research in Veterinary Science 12 : 183-184.

Ashdown, R.R. and Pearson, H. (1973a) Anatomical and experimental studies on eversion of the sheath and protrusion of the penis in the bull. Research in Veterinary Science 15 : 13-24.

Ashdown, R.R. and Pearson, H. (1973b) Studies on corkscrew penis in the bull. Veterinary Record 93 : 30-35.

Ashdown, R.R., Ricketts, S.W. and Wardley, R.C. (1968) The fibrous architecture of the integumentary coverings of the bovine penis. Journal of Anatomy 103 : 567-572.

Beckett, S.D., Walker, D.F., Hudson, R.S., Reynolds, T.M. and Vachou, R.I. (1974) Corpus cavernosum penis pressure and penile muscle activity in the bull during coitus. American Journal of Veterinary Research 35 : 761-764.

Beckett, S.D., Reynolds, T.M., Walker, D.F., Hudson, R.S. and Purohit, R.C. (1974) Experimentally induced rupture of corpus cavernosum penis of the bull. American Journal of Veterinary Research 35 : 765-767.

Bellenger, C.R. (1971) A comparison of certain parameters of the penis and prepuce in various breeds of cattle. Research in Veterinary Science 12 : 299-304.

Blockey, M.A. de B and Taylor, E.G. (1984) Observations on spiral deviation of the penis in beef bulls. Australian Veterinary Journal 61: 141-145.

Bouters, R., Vandeplassche, M., Forent, A. and Devos, (1960) De ulcereuse balanoposthitis bij fokstieren. Vlaams Diergeneeskunde Tijdschrift 29 : 171-186.

Boyd, C.L. and Henselka, D.V. (1972) Implantation of a silicone prosthesis for correction of bovine penile deviation. Journal of the American Veterinary Medical Association 161 : 275-277.

Carroll, E.J., Aanes, W.A. and Ball, L. (1964) Persistent penile frenulum in bulls. Journal of the American Veterinary Medical Association 144 : 747-749.

De Groot, R. and Numans, S.R. (1946) Over de erfelijkeid der impotentia coeundi bij stieren. Vlaams Diergeneeskunde Tijdschrift 71 : 372-379.

Doherty, M.L. (1985) Outbreak of posthitis in grazing wethers in Scotland. Veterinary Record 116: 372-375.

Formston, C. (1953) Fibropapillomatosis in cattle with special reference to the external genitalia of the bull. British Veterinary Jounral 109 : 244-248.

Glossop, C.E. and Ashdown, R.R. (1986) Cavernosography and differential diagnosis of impotence in the bull. Veterinary Record 118: 357-360.

Hofmeyr, C.F.B. (1967, a,b,c) Surgery of bovine impotentia coeundi I, II and III. Journal of the South African Veterinary Medical Association 38 : 275, 395, 399.

Hofmeyr, C.F.B. (1968, a,b,c) Surgery of bovine impotentia coeundi IV, V and VI. Journal of the South African Veterinary Medical Association 39 : (1) : 17-25, (2) 93-101, (3) 3-16.

Hudson, R.S. (1971) Thrombus of the corpus cavernosum penis in a bull. Journal of the American Veterinary Medical Association 159 : 754-756.

Konig, H. (1961) Impotentia Coeundi nach Penisknickung beim Stier. Schweizer Archive Tierheilkunde 102 : 119-134.

Larsen, L.H. and Bellenger, C.R. (1971) Surgery of the prolapsed prepuce in the bull; its complications and dangers. Australian Veterinary Journal 47 : 349-357.

Lagos, F. and Fitzhugh, H.A. (1970) Factors influencing preputial prolapse in yearling bulls. Journal of Animal Science 30 : 949-952.

Long, S.E. and Dubra, C.R. (1972a) Ulcerative lesions in bulls. Veterinary Record 90 : 15-16.

Long, S.E. and Hignett, P. (1970a) Preputial eversion in the bull : a comparative study of prepuces from bulls which evert and those which do not. Veterinary Record 86 : 161-164.

McEntee, K. (1969) In Pathology of Domestic Animals. 2nd Edition Ed. by Jubb, K.V. and Kennedy, P.D. Academic Press New York and London.

Metcalf, F.L. (1965) Treatment for haematoma of the bovine penis. Journal of the American Veterinary Medical Association 147 : 1319-1320.

Meredith, M.J. (1970) Bacterial content of semen collected by artificial vagina from bulls that erect the preputial epithelium. Veterinary Record 87 : 122-124.

Milne, F.J. (1954) Penile and preputial problems in the bull. Journal of the American Veterinary Medical Association 124 : 6-11.

Nickel, R., Schummer, A., and Sieferle, E. (1973) The Viscera of the Domestic Animals. Paul Parey, Berlin.

Noordsy, J.L., Trotter, D.M., Carnaham, D.L. and Vest Weber, J.G. (1970) Etiology of haematomoa of the penis in beef bulls - a clinical survey. Proceedings of the Annual Conference on Cattle Diseases. American Association of Bovine Practitioners. 6 pp. 333-338.

Olson, D., Segre, D., and Skidmore, L.V. (1960) Further observations on immunity to bovine cutaneous fibropapillomas. American Journal of Veterinary Research 21 : 233-242.

Pearson, H. (1972) Surgery of the male genital tract in cattle : a review of 121 cases. Veterinary Record 91 : 498-509.

Pearson, H. (1972) Penile neoplasia in bulls. Veterinary Annual 17: 40-43.

Roberts, S.J. (1971) Veterinary Obstetrics and Genital Diseases. Published by Author, Ithaca, New York.

Seidel, G.E. and Foote, R.H. (1967) Motion picture analysis of ejaculation in the bull. Journal of Reproduction and Fertility 20 : 313-317.

Snowden, W.A. (1965) The IBR-IPV virus : reaction to infection and intermittent recovery of virus from experimentally infected cattle. Australian Veterinary Journal 41 : 135-142.

Straub, O.C. (1970) Quoted by Roberts, S.J. (1971).

Vandeplassche, M., Bouckaert, J.H., Oyaert, W. and Bouters, R. (1963) Results of conservative and surgical treatment for fracture of the penis of Bulls. Proceedings of the XVIIth World Veterinary Congress. Hanover, Vol. 2 pp. 1135-1138.

Walker, D.F. (1964) Deviations of the bovine penis. Journal of the American Veterinary Medical Association 145 : 677-682.

Walker, D.F. (1970) Preputial disorders and deviations of the penis in the bull. Proceedings of the Annual Conference on Cattle Diseases. American Association of Bovine Practitioners. 40 pp. 322-326.

Walker, D.F. and Vaughan, J.T. (1980) Bovine and Equine Urogenital Surgery. Lea and Febiger, Philadelphia.

Wolfe, D.F., Hudson, R.S. and Walker, D.F. (1983) Common penile and preputial problems of bulls. Compendium on Continuing Education For the Practising Veterinarian. 5: S447-S456.

Young, S.A., Hudson, R.S. and Walker, D.F. (1977) Impotence in bulls due to vascular shunts from the corpus cavernosum penis. Journal of the American Veterinary Medical Association 171 : 643-648.

Chapter Six
THE PREPARATION OF TEASER MALES

A. INDICATIONS AND PRE-OPERATIVE CONSIDERATIONS

I INTRODUCTION

Teaser males are needed to pick out female animals on heat without inseminating them. The most common use is in herds or flocks where it is desired to use Artificial Insemination and teaser males are the most satisfactory way of diagnosing heat. Sometimes teaser males are used to reduce the work load on a valuable stud animal so that the latter has only to serve the females on heat and not pick them out. A third use is to bring a group of females into heat roughly simultaneously by psychological means - thus if a vasectomised ram is introduced into a flock just before the normal breeding season begins, a higher proportion of ewes will begin to have oestrous cycles sooner than if the ram were not so introduced.

II CHOICE OF METHOD

Generally speaking, some surgical method of rendering the teaser male incapable of begetting progeny is necessary as the teaser male runs loose with the female, but where hand teasing and mating is the rule, as in most horse-breeding establishments, the teaser male remains intact and is separated physically by a gate or fence from the females as well as being manually restrained.

Of the surgical procedures available for producing teasers, vasectomy is the operation of choice for preparing teaser rams (Weaver, 1967), boars (Godke and others, 1979) and stallions (Selway and others, 1977). Other options (see page 123) have disadvantages, particularly as regards proven, long term effectiveness. In the U.K., legislation states that vasectomy may only be carried out by a qualified veterinary surgeon.

There is concern that vasectomised bulls can still transmit venereal disease and there is, therefore, a general unwillingness to use them, even in closed herds. In the bull, therefore, operations on the penis have been widely used to prepare teaser animals. The penis itself may be amputated (so that it is too short to allow intromission), or it may be deviated (so that at erection it passes far too lateral to the female for intromission to be possible), or it may be retroflexed (i.e. deviated backwards so that intromission is impossible). Deviation is now the preferred procedure. In the U.K., however, operations to prevent the bull serving are considered an unacceptable mutilation and they are banned by law.

It is often stated that vasectomised bulls lose their libido more quickly than bulls with a deviated or amputated penis but many of these statements can be traced back to one source. In any case, statements are also made that the contrary is true. It is probable that management factors are important in determining how long a bull will retain effectiveness and more work needs to be done on this aspect of teaser animals.

III SELECTION OF ANIMALS

It is better to operate on young puberal animals than on older failed stud beasts as the former are usually easier to handle and manage, and are relatively easier to operate on and there is less risk of them having previously acquired transmissible disease. It is also much easier to operate on immature males than on mature animals and the prepuberal animal is easier than the puberal animal.

The size of the teaser male should be approximately that of the females with which he is to be used. Therefore, if puberal animals are to be used for teasers, their future growth should be taken into account.

It seems preferable (see page 124) to have at least two males operating at a time with a similar number resting and for there to be 1 male to 25-50 females. The number of males it will be necessary to provide will, therefore, have to be calculated.

In selecting bulls for penile deviation, it is important not to operate on pre-puberal animals as these may later learn to serve normally.

B. TECHNIQUES

I VASECTOMY

(a) <u>Pre-operative Considerations</u>

In order to reduce pre-operative contamination, prolonged yarding before operation should be avoided and it is preferable that the male should have been bedded on clean straw for at least 24 hours prior to surgery.

(b) <u>Anaesthesia and Restraint</u>

Although general anaesthesia is often recommended it is not always practical and in the bull carries with it a risk which may not be acceptable. Sedation with xylazine (note that young bulls are susceptable to xylazine and smaller doses than normal will be effective) and local anaesthesia of skin and spermatic sac are effective. There is a slight risk of injecting local anaesthetic into the pampiniform plexus but this can be obviated by drawing back on the syringe prior to injection. The ram is best restrained in the sitting position if sedation is employed or in dorsal recumbency if general anaesthesia is used. Bulls are best restrained in lateral recumbency if general anaesthesia is used with the uppermost hind leg held well forwards as for castration in the horse (see Appendix I. Fig. 2). Stallions must be given a general anaesthetic.

(c) <u>Operative Technique</u>

The area of the neck of the scrotum is prepared for surgery. A vertical skin incision of about 4 cm in the ram and 10 cm in the bull is made on the cranial surface of the neck of the scrotum to the left of the mid line over the neck of the left spermatic sac. By blunt dissection the left spermatic sac is freed and brought outside the skin incision where it may be held in place by a pair of haemostats behind it.

It is also possible to approach the spermatic sac through skin incisions made laterally or caudally in the neck of the scrotum but in the ruminant the cranial approach is easy, gives access to the sac over its longest surface and is least likely to undergo post-operative trauma or infection.

Selway and others (1977) have described an approach over the caudomedial aspect of the scrotum of the stallion - this is rather easier than using the short scrotal neck.

The deferent duct lies medially within the spermatic sac so that it is necessary to roll the sac outwards. It is usually possible to identify the duct within the sac by its hard solid texture. (Fig. 1:3 shows a cross-section of the neck of the sac). A nick is made in the vaginal tunic over the duct which then usually pops out - a spey hook helps to hold the duct out. The sac should not be incised <u>until</u> the duct is identified and is held firmly below the point of incision. To incise the sac and then search for the duct is to invite a blood bath. In the pig, the external spermatic fascia is substantial (see also page 60) and will need to be incised before access to the spermatic sac can be obtained. In this species, the sac is most readily located where it runs between the thighs from the sub-anal scrotum to the inguinal canal. Identification as the duct is confirmed by palpation and by the presence of a small artery and vein which run along the duct. A portion of the duct is exteriorised and a length not less than 3 cm (5 cm in the stallion and bull) and preferably longer is removed to reduce the risk of re-anastomosis. The ends of the sectioned duct are ligated with fine absorbable suture material and returned to the lumen of the sac. The hole in the sac need not be sutured. Any significant deadspace created by dissection of the sac is closed but this is not usually necessary. The skin is sutured.

The whole procedure from skin incision to skin closure is repeated to the right of the mid line.

The pieces removed should be sent for independent histological confirmation that they were, in fact, deferent duct. This will not only satisfy the surgeon that he has removed the correct tissue but will also be useful evidence in the case of eventual litigation. As an immediate test, the contents of the excised duct may be squeezed onto a microscopic slide and examined for the presence of sperm.

II PENILE DEVIATION

(a) <u>Pro-operative Considerations</u>

This operation can be carried out in a number of ways but the technique described below is basically that of Royes and Bivin (1973), modified by Copland (1975). The method takes advantage of the fact that the penile blood supply (see Chapter Five) arises entirely from the root of the penis (the internal pudendal vessels) and not from the peripheral tissues. It should not be done in pre-puberal or young puberal bulls as these animals seem to be able to learn how to serve with a deviated penis. The bull is placed in dorsal recumbency under deep narcosis or, preferably, general anaesthesia. An area extending from 20 cm cranial to the preputial orifice to the scrotum and laterally out to the fold of the flank is prepared for surgery. If the surgeon is right-handed it is preferable that the left side of the ventral body wall is prepared.

(b) **Operative Technique**

Skin incisions are made in the mid-line as shown in Fig. 6:1 and the triangular areas cranial and caudal to the preputial orifice are removed.

The penis and its blood vessels and the preputial orifice with its surrounding skin are dissected free of the underlying tissues. This entails severance of the cranial muscles of the prepuce cranially and the preputial branches of the caudal abdominal vessels on the side opposite to the deviation. (The vessels on the side of the deviation should be preserved if at all possible but this is difficult as they go into spasm and can only be identified with difficulty. If they are severed, preputal oedama will develop post-operatively but this usually resolves spontaneously.)

A section of skin corresponding in size to the circle round the preputial orifice is removed from the ventral body wall laterally, just medial to and slightly cranial to the fold of the flank. Its exact location is determined by placement of the severed preputial orifice. The angle between mid-line and the new angle for the penis should be between 35° and 40°.

A pair of scissors is used to make a sub-cutaneous tunnel from the new site for the preputial orifice to the scrotum - the tunnel should be directed as far caudally as possible.

The preputial orifice and penis are snared and then pulled to their new location. After ensuring that no gross angulation or distortion of the penis occurs, the preputial orifice is sutured into its new location and the mid-line skin incision is also repaired.

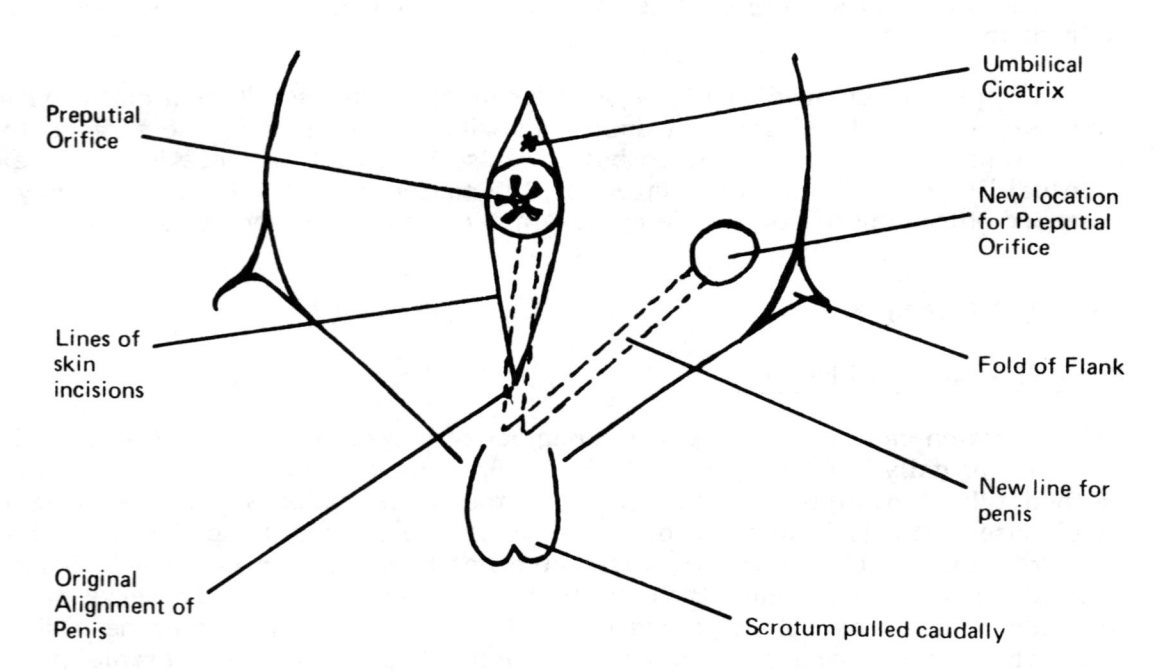

Fig. 6:1 Diagram showing Penile Deviation Operation.

IV OTHER TECHNIQUES

(a) Cauda Epididymectomy

In this technique the epididymal tail is resected where it lies ventral to the testis. The technique has a high failure rate not only because the animal takes considerably longer than a vasectomised male to become infertile (Moller, 1971) but also because of the eventual formation of a spermatocoel which allows sperm to pass from epididymal head to deferential duct (van Rensberg, McFarlane and van Rensberg, 1963). The technique can only be applied where immature ram lambs are prepared for use the following season and sent for slaughter within 12 months of birth. Details of the technique are given in Dun (1960) and Smith and Fletcher (1968).

(b) Injection of Sclerosing Agents into Epididymal Tail

The success of this technique depends upon the agent used and work remains to be done on the long term efficacy (measured in years) of the agents used. For instance, Morrant (1959) showed that the injection of a mineral oil called Dondren was followed by aspermia in rams but recanalisation occurred in three months. Pearson and others (1980) have described the injection of 2.5ml or 5ml of a 3% solution of chlorhexidine gluconate in 50% dimethyl sulphoxide in water into the epididymal tail percutaneously through a 25 x 1.1mm needle. Aspermia of at least twelve months duration resulted. Mercy and others (1985) have injected 0.75ml of 102 formaldehyde in alcohol into ram epididymides but some of these rams still ejaculatated live sperm at 48 days, although the authors claim they would be infertile. Similar problems have been observed in stallions injeted with chlorhexidine gluconate/DMSO mixtures (Squires and others, 1978).

(c) Induced Cryptorchidism by the Use of Rubber Rings

The testes are pushed upwards from the scrotum to an inguinal position and retained there by placing an 'Elastrator' rubber ring round the neck of the scrotum. The cryptorchid testis is sterile but the animal retains his libido. The procedure is easiest to carry out on pre-puberal animals but this means there is a considerable time before they can be used. In puberal animals the failure rate is said to be unacceptably high.

(d) Closure of the Preputial Orifice

In this technique, the preputial orifice is closed and a fistula created several cms caudal between the preputial cavity and the outside. Urine scald is common, and the technique is not, therefore, to be recommended.

(e) Use of an apron and other mechanical devices

An apron may be fitted round the male which covers the preputial orifice and a considerable distance round this. The idea is that this apron should keep the penis close to the male's body and so prevent intromission. In use, however, the apron becomes twisted or caught on fences, etc. and no longer does its job effectively.

Other mechanical devices some of which are fitted into and anchored in the preputial cavity and prevent protrusion have been described. The rate of sepsis is unacceptably high and the useful life of many of the animals so prepared exceedingly short (see Wenkoff, 1975).

(f) Corpus cavernosal block

Walker and Vaughan (1980) have described the injection of 10ml of a 1:1 mixture of freshly prepared Technovit into the corpus cavernosum of the bull at the distal bend of the sigmoid flexure, just behind the serotum to prevent erection (see page 120 on Impotence). In theory, the technique should be highly effective.

C. POST-OPERATIVE CONSIDERATIONS

I INTERVAL TO USE

Dunlop, Moule and Southcott (1963) observed that although sperm might be present in the ejaculate of vasectomised rams for a considerable time after vasectomy, these sperm were dead and, therefore, the animal is infertile. The concluded that it was possible to put vasectomised rams to use safely one week after vasectomy. Similar observations have been made for bulls (Nair, Nair and Jalaludeen, 1979).

Selway and others (1977) found live sperm up to 4 weeks post-vasectomy in stallions but none at 6 weeks.

Males prepared with penile deviation should be left for at least a month after operation for complete healing to occur before they are used.

II MANAGEMENT

It is essential that the farmer recognises that teaser animals are males and should be treated as if they were entires - it is especially dangerous with bulls to do otherwise! Moreover, if the animal is to maintain his interest and libido, he should be fed and cared for as though he were the stud male - he should not be expected to survive on half rations or to serve twice as many females as a stud.

In order to maintain interest and activity, it is as well to keep to the following schedule:

(i) The teaser males should be allowed to run with the females prior to the breeding season.

(ii) The teaser males should be provided with a marker so that mounted females can be picked out. The colour of the marker should be changed at intervals not exceeding 4 days less than the normal oestrous cycle length for the female of the species.

(iii) All teaser bulls should be dehorned and have their noses rung.

(iv) Teaser males should be worked with at least one other male at a stock density of not less than 1 male/50 females. Two males will compete with each other and if two bulls are used there is also less likelihood of selection by one bull resulting in a female on heat being missed. There is little danger of fighting unless the actual stocking density is too high (it might be so in continually yarded animals). Overworking of teaser animals is a common cause of their becoming ineffective. The teaser male's interest may be stimulated by allowing him a week of rest alternating with a week of activity but it is not always practicable to remove them in this way.

(v) Freshly marked females should be removed for insemination or hand-mating where it is practicable to do so, in order to divert the teaser male's attention to other females.

D. REFERENCES

Dun, R.B. (1960) Artificial insemination in sheep IV. Sterilisation of rams. Australian Veterinary 36 : 437-439.

Dunlop, A.A., Moule, G.R. and Southcott, W.M. (1963) Spermatozoa in the ejaculates of vasectomised rams. Australian Veterinary Journal 39 : 46-48.

Godke, R.A., Lambeth, V.A., Kreider, J.L. and Root, R.G. (1979) A simplified technique of vasectomy for heat-check boars. Veterinary Medicine 74 : 1027-1029.

Mercy, A.R., Peet, R.L., Johnson, T., Cousins, D.V., Robertson, G.M., Batey, R.G. and McKenzie, D.P. (1985) Evaluation of a non-surgical technique for sterilising rams. Australian Veterinary Journal 62 : 350-352.

Moller, K. (1971) Sterilisation of bulls. New Zealand Veterinary Journal 19 : 185-187.

Morrant, A.J. (1959) Sterilisation of teaser rams. Australian Veterinary Journal 35 : 368.

Pearson, H., Arthur, G.H., Rosevink, B. and Kakati, (1980) Ligation and sclerosis of the epididymis in the bull. Veterinary Record 107 : 285-287.

Nair, K.P., Nair, K.N.P. and Jalaludeen, A.M. (1979) Disappearance of spermatozoa from the ejaculate of bulls following vasectomy/caudectomy. Indian Journal of Animal Science 49 : 1009-1014.

Royes, B.A.P. and Bivin, W.S. (1973) Surgical displacement of the penis in the Bull. Journal of the American Veterinary Medical Association 163 : 56-57.

Selway, S.J., Kenney, R.M., Bergman, R.V., Greenhough, G.R., Cooper, W.L. and Ganjam, V.K. (1977) Field technique for vasectomy. Proceedings of the Annual Convention of the American Association of Equine Practitioners 23 : 355-361.

Smith, J.S. and Fletcher, F.K. (1968) Vasectomy in the Ram. Veterinary Record 83 : 110.

Squires, E.L., Pineda, M.H., Seidel, G.E. and Pickett, B.W. (1978) Chemical vasectomy of stallions. Journal of Animal Science 47. Supplement pp. 391-392.

Van Rensberg, S.J., McFarlane, I.S. and Van Rensberg, S.W.J. (1963) Sterilisation of teaser male ruminants -the reliability of surgical methods. Journal of the South African Veterinary Medical Association 24 : 249-253.

Walker, D.F. and Vaughan, J.T. (1980) Bovine and Equine Urogenital Surgery. Lea and Febiger, Philadelphia.

Wenkoff, M.S. (1975) Problems associated with teaser bulls prepared by the Pen-o-block method. Canadian Veterinary Journal 16 : 181-186.

Weaver, A.D. (1967) Vasectomy in the ram. Veterinarian 4 : 155-159.

Chapter Seven
VAGINAL PROLAPSE

Prolapse of the vagina occurs commonly in sheep, less commonly in cattle, rarely in pigs and probably never in horses. It occurs all over the world under a variety of conditions although it is most common in late pregnancy. It has a whole host of colloquial names including 'bearing trouble', 'red bag', 'showing the rede', throwing the rose', pushing out the button' and, as Edgar (1952) observed, 'no doubt by other picturesque local names'.

A. ANATOMY AND PATHOLOGY

The normal anatomy of the pelvic organs of the ruminants is represented in Fig. 7:1. Although all the structures described below are mentioned in standard anatomy texts, the most detailed account of the relevant anatomy will be found in Bassett (1965). The mare is not considered further.

I   PELVIC LIGAMENTS

The viscera of the pelvis are supported by the following ligaments:-

(a)   Mesorectum

This attaches the rectum to the dorsal body wall. It is not very deep dorso-ventrally and merges cranially with the mesocolon.

(b)   Broad Ligaments

These paired triangular structures arise from the lateral wall of the pelvic canal and the caudal body wall ventral to the tuber coxae and attach to the ventro lateral surface of the uterine horns. They contain, especially after several pregnancies, substantial amounts of smooth muscle fibre in the form of flat strands or sheets. These fibres are continuous with the longitudinal fibres of the uterus and raise the possibility that the uterus is actively supported, especially during pregnancy.

The most cranial point of origin of the ligaments is at the level of the tuber coxae so that in the non-pregnant animal the cranial borders of the ligaments are directed caudo-ventrally whilst in pregnancy they run somewhat more directly ventrally, the enlarged uterus spilling forwards between them. The absence in ruminants of mesometrial origins from the lumbar area makes the uterus more easily displaced caudally and also more likely to twist. A clear distinction must be made between the vagina (cranial to the urethral orifice) and the vestibule (caudal to the urethral orifice). The latter is firmly attached to the pelvic floor and anal ligaments, the former is only loosely attached to the pelvic floor and is, therefore, readily displaced.

(c) Ligaments of the Bladder

The paired lateral ligaments of the bladder arise from the lateral wall of the pelvis and pass medially to attach to the sides of the bladder. Their cranial boundaries are formed by the lateral ligaments of the bladder, the adult remains of the paired umbilical arteries. This boundary, therefore, runs from the internal iliac artery to the apex of the bladder. The single median ligament of the adult is small and runs between the ventral surface of the bladder and the pelvic floor (in fetal life it was the supporting fold of the urachus and extended to the umbilicus but this cranial area degenerates after birth).

In the cow, the neck of the bladder and urethra are united with the ventral wall of the vagina by tough connective tissue.

II PELVIC DIVERTICULA

Three diverticula or pouches of the peritoneal cavity separate the rectum, genital tract and bladder from each other and the body wall to varying degrees, each pouch having a name appropriate to the structures it separates.

(a) The Recto-Genital Pouch

This is the most dorsal of the three. It is bounded ventrally by the uterus, cervix and cranial vagina and laterally by the broad ligaments. Dorsally it is split into two parts by the rectum suspended in its mesorectum, thus creating what are sometimes called the sacro-rectal pouches. The pouch extends deeply into the pelvic canal, to about the level of the 1st or 2nd coccygeal vertebrae. Thus the most cranial third of the dorsal vaginal wall is separated from the peritoneal cavity only by peritoneum.

(b) The Vesico-Genital Pouch

This is bounded dorsally by the uterus and broad ligaments and ventrally by the bladder and its lateral ligaments. It does not extend quite as far caudally as the recto-genital pouch.

(c) The Vesico-Pubic Pouch

This is bounded dorsally by the bladder and ventrally by the pelvic floor and body wall, and is split by the median ligament of the bladder. It extends into the pelvic canal about the same extent as the vesico-genital pouch.

The consequent mobility of parts of the reproductive tract and the bladder and the propensity of their ligamentous supports to stretch make it possible for the bladder and/or the uterus and vagina proper to pass caudally to such an extent that they may come to lie beyond the vulval lips.

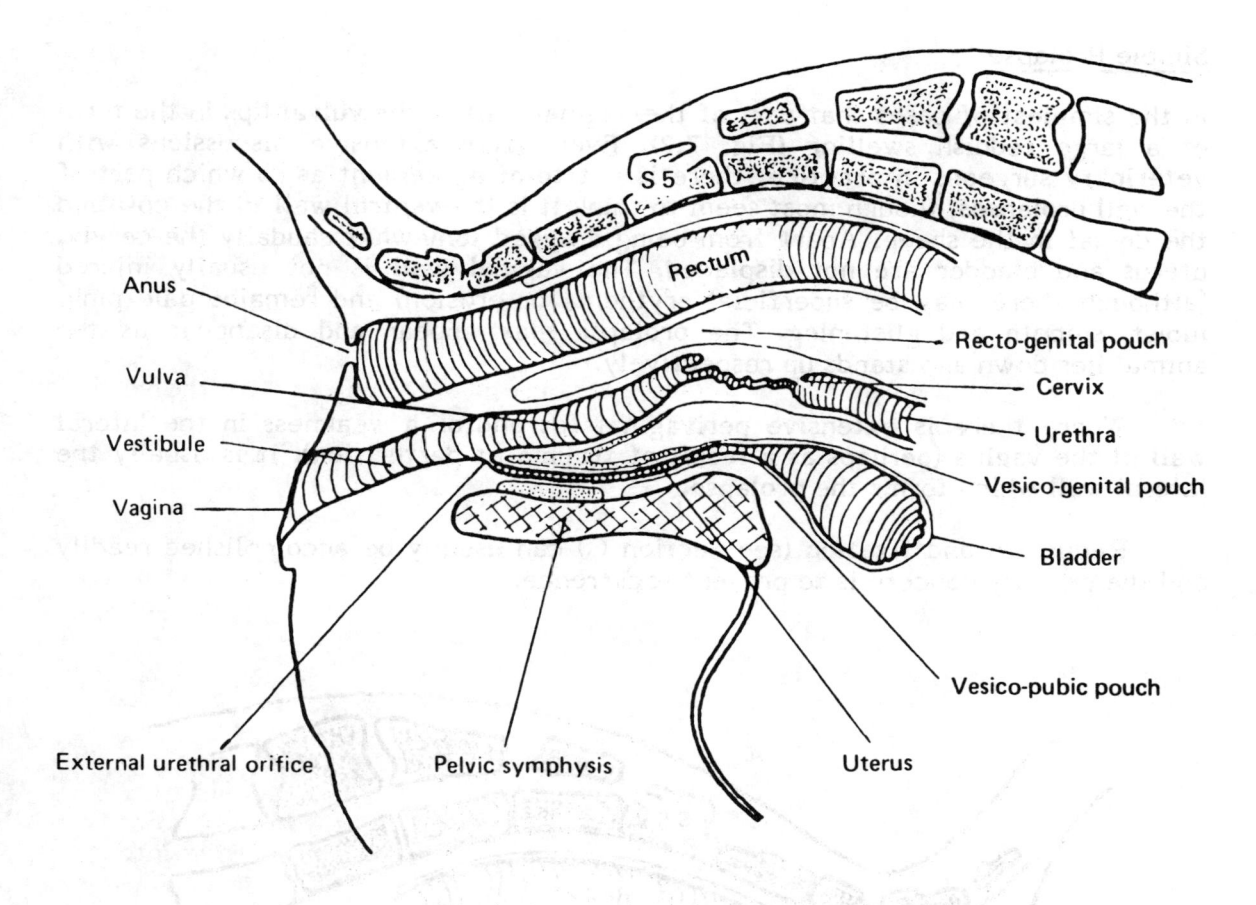

Fig. 7:1 Sagittal section through the pelvic region of a normal ewe showing the peritoneal pouches.

III PROLAPSE

Vaginal prolapse may, therefore, be said to occur when a part or the whole of the vaginal wall is displaced in such a way that its mucosa is visible at the vulval lips. Associated with the vaginal prolapse there is usually displacement of other organs as well, viz. bladder, uterus and cervix, intestines. The vestibule (from urethral opening to vulva) is firmly attached to the pelvic floor and anus and is, therefore, rarely displaced.

Varying degrees of prolapse may be recognised according to:

(i)   the size of the prolapse;

(ii) the involvement of organs other than the vagina;

(iii) the length of time for which the prolapse has been present.

(a) <u>Simple Prolapse</u>

In the simplest of cases that part of the vaginal wall at the vulval lips in the form of a large reddish swelling (Fig. 7:2). Even after extensive discussions with veterinary surgeons in practice, there is not total agreement as to which part of the wall prolapses, though most seem to think it is the ventral wall in the cow and the dorsal in the sheep. Apart from being dragged somewhat caudally the cervix, uterus and bladder are not displaced. The vaginal wall is not usually injured (although there may be superficial erythema or erosion) and remains pale pink, moist, smooth and glistening. The prolapse may appear and disappear as the animal lies down and stands up respectively.

Where there is extensive perivaginal fat and/or a weakness in the lateral wall of the vagina (perhaps as a result of trauma at parturition), it is usually the lateral wall which forms the prolapse.

Reduction and fixation (see Section C) can usually be accomplished readily and the primary concern is to prevent recurrence.

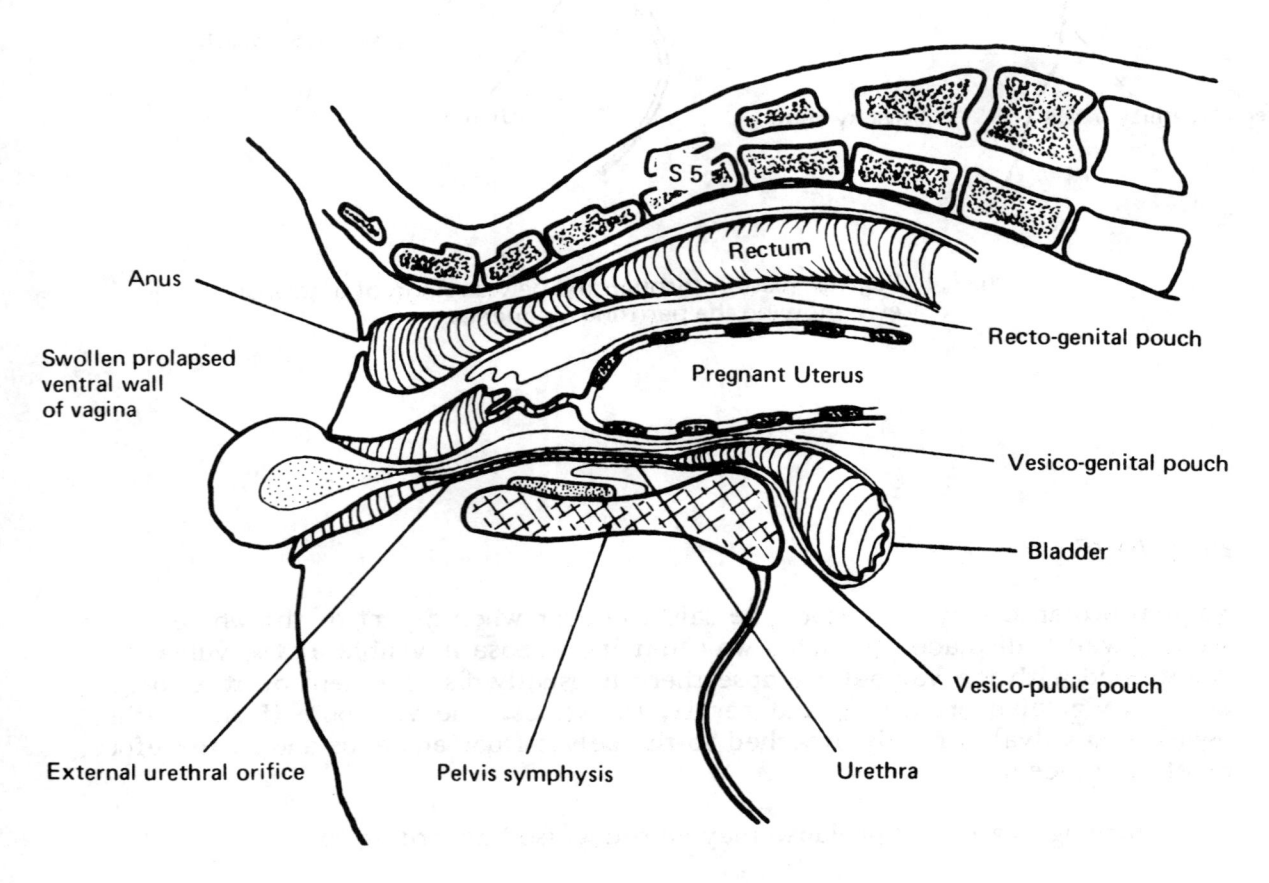

Fig. 7:2 Sagittal section through the pelvic region of a ewe with simple vaginal prolapse.

(b) **Moderate Prolapse**

The next stage occurs when bladder (Fig. 7:3) or intestine become involved in the prolapse and get trapped in the pelvis. This may happen because:

(i) the vaginal prolapse has become so great and the vesico-genital pouch is greatly increased in size allowing viscera to enter it;
or

(ii) exposure of the prolapsed tissue results in oedema and passive congestion.

The effect is to make the animal strain more at the foreign mass in the pelvis and this in turn increases both the size of the prolapse and the involvement of other organs. The animal becomes listless and uneasy, grazes only a little and separates herself from the herd or flock. It is unusual for intestine to be so badly trapped that symptoms of intestinal obstruction develop, and strangulation ensues. It is also unusual in the early stages for urination to be completely hindered, although because of the tenesmus, the owner may believe the animal has difficulty urinating. Soon, however, urination does become impossible and the animal may die of rupture of the bladder and uraemia.

Spontaneous reduction is now unlikely and the exposed vaginal mucosa becomes more severely ulcerated and dried. Its colour changes from pale pink through deep red to blue. In severe cases it may become black and actually perforate with necrosis leading to peritonitis or to a rapidly fatal haemorrhage.

Reduction is somewhat more difficult than in the simple cases and must be accompanied not only by steps to prevent recurrence but also by control of the vaginitis. If a ruptured bladder or peritonitis is present the animal should be slaughtered.

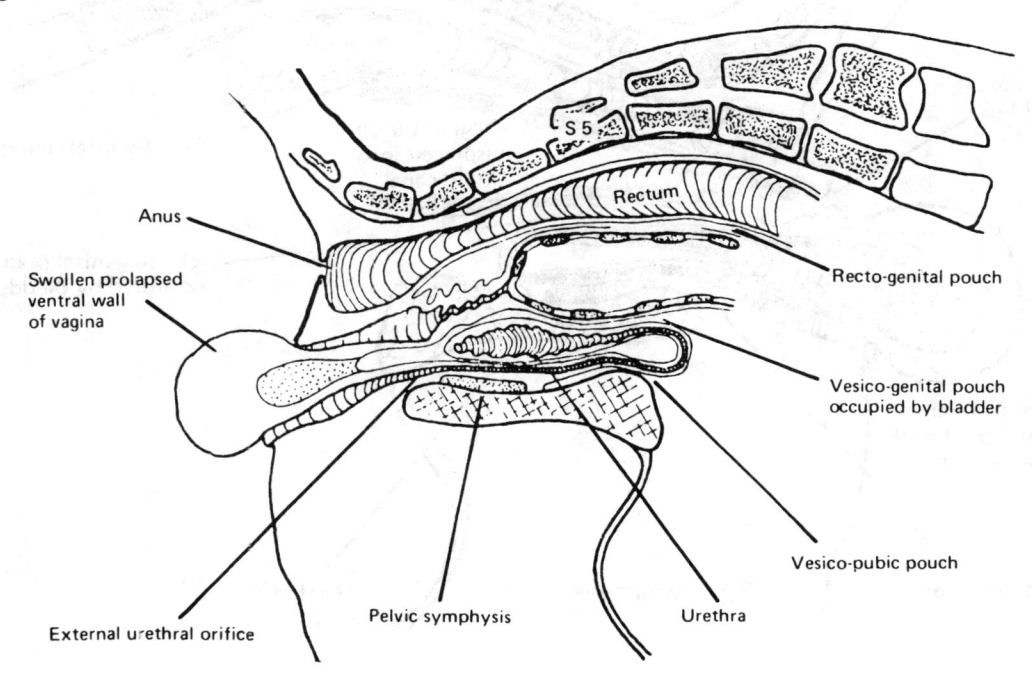

Fig. 7:3 Sagittal section through the pelvic region of a ewe with moderate vaginal prolapse.

(c) <u>Severe Prolapse</u>

The third and most severe stage occurs when the uterus and cervix have been pushed so far caudally that the cervix appears at the vulval lips (Fig. 7:4). Once again, extensive discussion with veterinary surgeons in practice has failed to produce conclusive evidence that this Figure is exactly right. There is some consensus that in the cow it may only be cervix which appears at the vulval lips. The normal sequel is bacterial liquefaction of the cervical seal and the establishment of infection inside the uterus. Fetal death and abortion are, therefore, possible sequelae and may occur several days after correction. The fetus almost invariably becomes emphysematous. Walker and Vaughan (1980) comment that in cattle under no circumstances should the operator attempt to dilate the prolapsed cervix and deliver the fetus in an attempt to avoid the difficulties of caesarean on an emphysematous fetus - they state that this invariably leads to cervical and uterine tearing. Nevertheless, the cervix can be dilated in some cases in cattle and many in sheep and the fetus delivered per vaginam. Other cases will need a caesarean operation.

There is an increased mortality in both affected ewes and in lambs born to affected ewes - Edgar's survey suggested that 33% of the lambs were born dead or died. Woodward and Quesenberry (1956) observed an 18% mortality in range cattle with the condition but commented that the number of fatalities was determined largely by adequacy and time of treatment - if a cow could be moved for treatment under relatively clean conditions the prognosis was good.

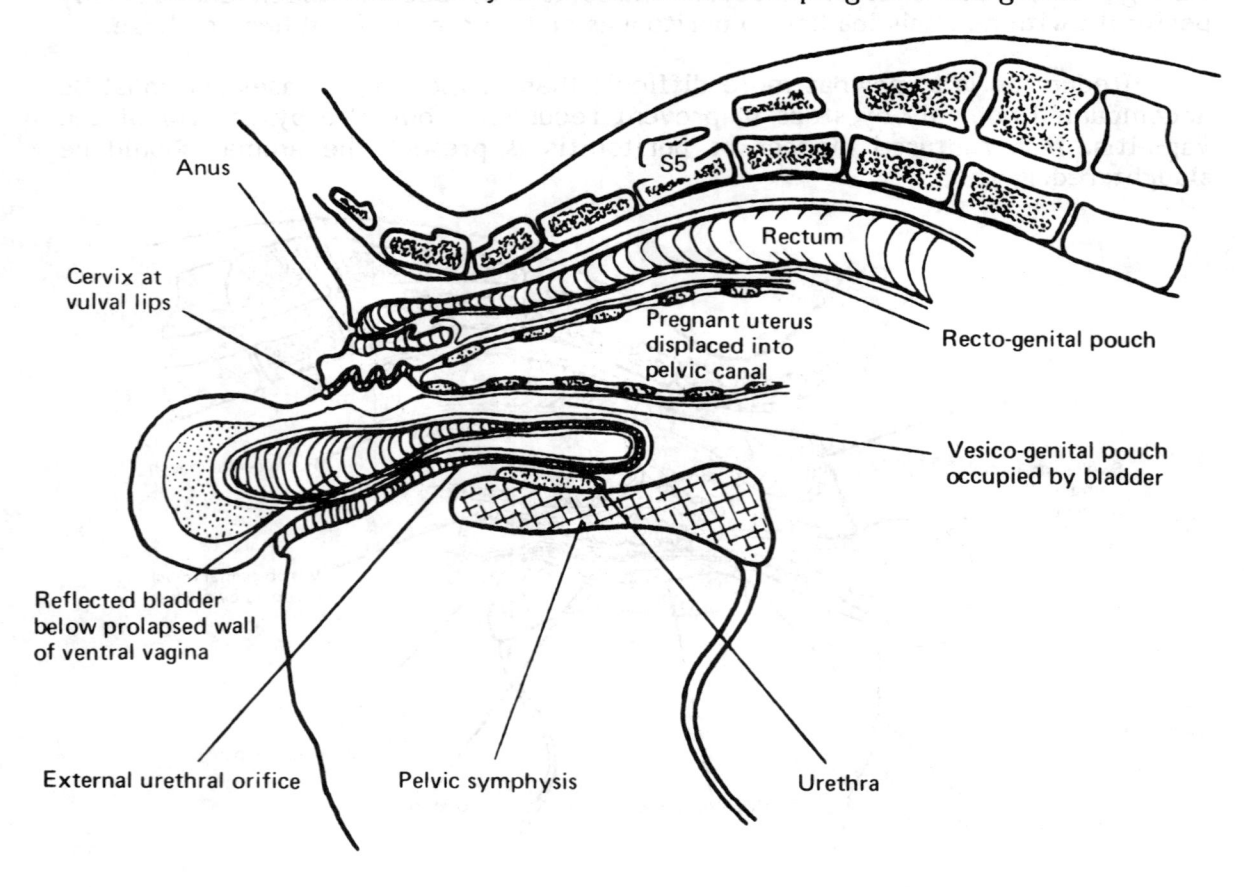

Fig. 7:4 Sagittal section through the pelvic region of a ewe with severe vaginal prolapse.

B. AETIOLOGY AND INCIDENCE

I INTRODUCTION

Most of the detailed investigations of vaginal prolapse have been carried out in sheep (Edgar, 1952; McLean 1956 et passim and Bassett 1955 et passim).

The condition occurs in sheep primarily during the last two weeks of pregnancy although in dry ewes and just after lambing. Up to 20% of sheep in a flock can be affected but the more usual rate is about 1-5%. It seems more common in Romney Marsh sheep in New Zealand than in, say, Cheviot, Merino and Ryeland breeds (Laing, 1949) and a high incidence is reported in Kerry Hill and Clun Forest in the U.K. (Arthur, 1975) but whether this is a real breed incidence or a reflection of management procedures is not established. Other factors which affect the incidence are discussed in the following pages.

In cattle, the condition occurs rarely in nymphomaniacal (see II below) and fat non-pregnant cows or heifers. It is most commonly seen in beef cattle, particularly in Herefords in the western ranges of the U.S.A. in the last two months of pregnancy. The incidence is rarely more than 1% of a herd. Woodward and Quesenberry (1956) concluded that there was a tendency for the condition to be inherited. The condition occurs rarely in pigs, usually accompanied by other signs and in animals fed on mouldy maize (see p.134).

McLean (1956) has pointed out that for prolapse to occur three conditions must be met:-

(i) The vagina must be in a state in which it can be easily everted. It is more likely to be in such a state if it is soft and relaxed with a potentially large lumen than when it is tense and firm and has a contracted lumen.

(ii) Prolapse will be easier if the vulvar and vestibular tissues are relaxed and will probably be impossible if these tissues cannot dilate.

(iii) Some force must operate to push the relaxed tissue out of its normal position.

II VAGINAL AND VULVAR RELAXATION

An annular vaginal constriction lies just cranial to the urethral orifice. This constriction is at the level of the end of the paramesonephric (Mullerian) duct and appears (Bassett, 1965) to be an integral part of the vaginal wall with very much more smooth muscle than elsewhere in the vagina. It is most pronounced in maiden ewes but is present in all ewes. Bassett and Philips (1955) have shown that the circumference of the vagina of the normal sheep at this point is considerably greater in late pregnancy than at any other time and that the potential displacement of the reproductive tract was also greatest at this time. They also showed that ewes which developed prolapses tended to have a greater potential for displacement of the reproductive tract and for vaginal distensibility than control ewes which did not develop prolapses. McLean (1956) has shown that the dilatability of the ewe's vagina increases dramatically during pregnancy, reverting to the non-pregnant state within 14 days of lambing, although primiparous ewes never return quite as low as in their virginal state. McLean also observed similar changes in vestibular relaxation but these were not as dramatic.

These changes are probably due to connective tissue changes brought about by hormone changes and oestrogens have been particularly implicated. Cows or heifers given large doses of oestrogen can develop vaginal prolapse (Folley and Malpress, 1944) and Garm (1949) has recorded cases of vaginal prolapse in nymphomaniacal cows, although such cases must be considered the exception rather than the norm. Edgar (1952), however, adminstered oestrogens in sheep in late pregnancy and observed no spontaneous cases of prolapse. Bennetts (1944) has reported prolapse of the uterus, but (importantly and contrary to widely held views) not the vagina, in sheep ingesting highly active oestrogenic substances during prolonged grazing of subterranean clover pastures in Australia. A recent study (Sobiraj, Busse, Gips and Bostedt, 1986) found increased concentrations of plasma oestradiol in ewes with vaginal prolapse in late pregnancy in comparison with controls, but the differences were slight. These same authors found lower calcium concentrations in affected sheep, but no clinical signs of hypocalcaemia were present.

Vaginal prolapse has been seen by several workers when mouldy maize was fed to non-pregnant sows and gilts and was associated with vulval enlargement and mammary growth (see McErlean, 1952). Bassett (1955) noted that ewes which had a history of prolapse tended to have deeper and broader pelves and, when pregnant, to have looser sacro-iliac joints than normal and Pandit, Gupta and Pattabiraman have made similar observations in buffaloes (1983). The author, moreover, has seen an 'outbreak' in extremely fat non-pregnant Charolais heifers and it was noticeable that the condition only occurred in animals with a deep pelvis.

III THE ROLE OF WEIGHT

Classically vaginal prolapse occurs in sheep in exceptionally good bodily condition, being fed on improved pastures or in seasons of good growth. It is also much more common in ewes carrying more than one lamb and in older ewes and cows (though a report from India suggests it occurs more commonly in buffaloes). As noted previously, it is primarily a condition of late pregnancy when the animal's weight is at its maximum. Such animals also tend to take less exercise than normal and this may be a contributing factor to overweight. However, prolapse can also be seen in both cows and sheep in poor condition in late pregnancy when they have been fed a high residue diet such as turnips. An important predisposing factor, therefore, seems to be abdominal distension (primarily the rumen) rather than good condition per se.

It has also been postulated that ewes in late pregnancy become sluggish and delay urination until a number of factors, including pressure from the heavy uterus on the bladder, render micturition difficult. They then strain more and a prolapse is initiated.

McLean and Claxton (1960) showed that any increase in live weight, whether associated with pregnancy or not, results in an increase in intravaginal pressure and that, conversely, emptying of the digestive tract by fasting leads to a decrease. Changes in intravaginal pressure may then be one of the factors which initiates prolapse.

McLean (1959) also showed that a high place of nutrition did not affect the degree of relaxation of the pelvic tissue - that was a consequence solely of the stage of pregnancy.

Thus in ewes with distended abdomens at the end of pregnancy, two factors - vaginal relaxation and increased intra-vaginal pressure - coincide.

III POSTURE IN RECUMBENCY

McLean (1957) observed that prolapses in New Zealand occurred more frequently in sheep in hill country than in lowland country and he noted that ewes in late pregnancy in hill country tended to lie with their head up-hill and their pelvis down-hill. McLean and Claxton (1960) were able to show that in sheep which adopted this tilted posture, the intra-vaginal pressure was increased above that in ewes lying in the horizontal position. Here, then, may be the extra and critical factor which tips fat ewes in late pregnancy over into prolapse - an increase in intravaginal pressure because they lie with their heads up-hill! This concept receives support from two clinical observations. Firstly Palsson (quoted by Edgar) reported that in Iceland, where pregnant ewes are housed, the incidence of the condition is reduced by allowing the rear part of the standing to become built up with dung. Secondly, it was observed in the Liverpool University Farm Practice that vaginal prolapse occurred in dry sows restrained in stalls whose design was such that the pelvis was considerably lower than the head (Faull, personal communication).

C. TREATMENT

I   REPLACEMENT

Physical replacement of the vagina is considerably facilitated by holding the ewe's hind quarters higher than her fore quarters (Fig. 7:5) or in the cow by low epidural anaesthesia.

Fig. 7:5 Showing one way of restraining a sheep for lambing or replace of prolapse vagina. The fore legs are tied by a calving rope and the sheep is slung by the hind legs round the operator's neck using a second calving rope.

The exposed surface is gently cleaned with a mild antiseptic solution at blood heat if possible. The displaced bladder may be emptied by lifting the prolapsed mass to release the kink in the urethra and then applying pressure. Persistent pressure, which may need to be quite considerable, replaces the prolapse. The prolapse should be kept in place by a hand until it is warm and until the ewe passes water.

II   PREVENTION

Laing (1945) suggests that where the condition is likely to occur and where the ewes appear lazy, the flock should be quietly mustered once daily and changed to a new pasture. Equally beneficial is a reduction in the roughage content of the diet.

III   TEMPORARY RETENTION

Bruere (1956) has described the successful treatment of cases occurring close to lambing by placing them temporarily in a crate 45 cm wide x 80 cm high x 120 cm long with a rear door for access and placing the rear end 10 cm higher than the front. Once the prolapse has been replaced, the ewe is left in the crate without food for 12-24 hrs. and then released. Relapses are rare.

The majority of authors, however, recommend some form of temporarily preventing recurrence of the prolapse so that the animal may continue pregnancy safely to term. The method used must be capable of being 'undone' or 'by-passed' at parturition so that the fetus may be delivered.

These techniques rely either upon making vulval and vestibular relaxation insufficient for prolapse to occur or upon preventing the vaginal wall being pushed far enough caudally to prolapse.

(a) Intravaginal Devices

A number of devices which are inserted into the vagina and held in place by strings or harnesses have been described, particularly for sheep. They include the following:

(i) a simple U-shaped piece of metal inserted blunt end first with the pointed ends left outside the vulva or bent outwards and tied to the fleece (see Fowler and Evans, 1957, Jones, 1958);

(ii) inflated inner case of small footballs (see Mayar, 1958);

(iii) a plastic tube with a cruciform base to which a special harness is attached ('Moffatt Prolapse Controller', Arnold and Sons).

(b) Vulval Truss

This method is illustrated in Fig. 7:6. the truss is made using bailer twine from the farm itself and it is possible for the ewe to lamb past it, a considerable advantage to a busy shepherd in a severe outbreak.

More sophisticated devices with canvas straps to go round the vulva are available commercially and may be more successful than twine in a severe case.

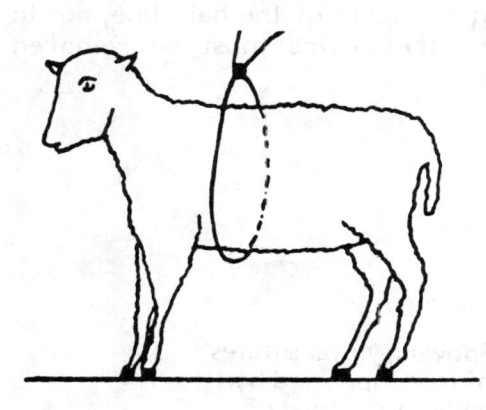

Fig. 7:6a Part the fleece and tie a loop of bailer twine tied fairly tightly round the chest just behind the shoulders.

Fig. 7:6b Two lengths of twine are attached to the chest loop and then knotted together over the lumbar spine.

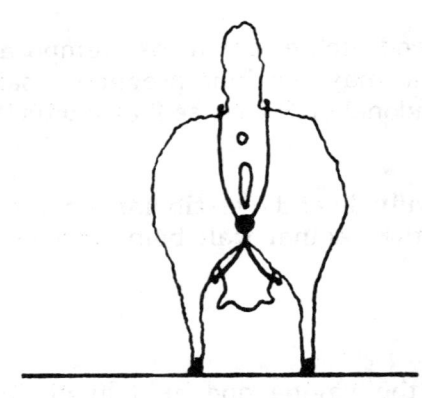

Fig. 7:6c The twines are taken down on either side of the tail, passed between the tuber ischii and the vulval lips and may again be knotted above or (as shown here) below the vulva or not at all.

Fig. 7:6d The twines are passed between the udder and thighs on either side and tied quite tightly to the knot over the spine with a bow, so that the tightness can be adjusted as required. The truss should be checked every few days for position, cutting and swelling.

(c) <u>Vulval Sutures or Clips</u>

As with intravaginal devices, a whole array of items and ideas have been employed to secure the vulval lips against the passage of the vagina (see Pierson, 1961).

These include the employment of mattress sutures, perhaps supported by quills or buttons (Fig. 7:7) or the use of ordinary safety pins or of special prolapse pins with a ball on each end. They are most often used in cattle but tend to produce necrosis of underlying tissue and may fall out if needed for long. If these techniques are used the device <u>must</u> be placed at the level of the hair-line, not in the hairless skin nearer the <u>vulva</u>. In all cases the device must be removed immediately prior to parturition.

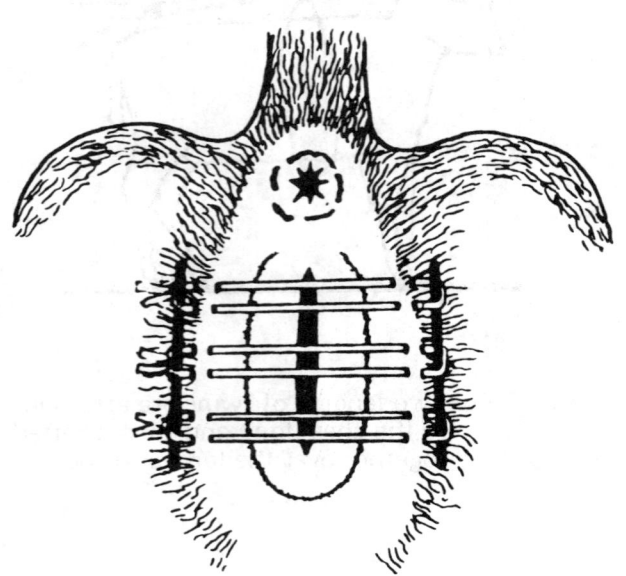

Fig. 7:7

Showing vulval sutures of tape supported by quills at the level of the hair-line.

(d) **Perivulval Sutures**

As an alternative to placing sutures across the vulval lips, the sutures may be placed round the vulva in the perivulval tissues.

Although a simple purse string technique is effective (see Pierson, 1961) the multiple entry points mean that the risk of infection gaining entry is high. The technique described by Buhner (see Bierschwal and DeBois, 1971) is preferred. The method uses a special needle (the Buhner needle) and about 2 feet of special tape (1 cm wide tubular, woven flattened synthetic material). The operation is all but impossible without the needle and, although umbilical tape can be used in place of the special tape, the latter is far superior.

A low epidural anaesthetic is given and the prolapse is reduced. The perineal and vulval areas are prepared for surgery - see Figs 7:8a - c.

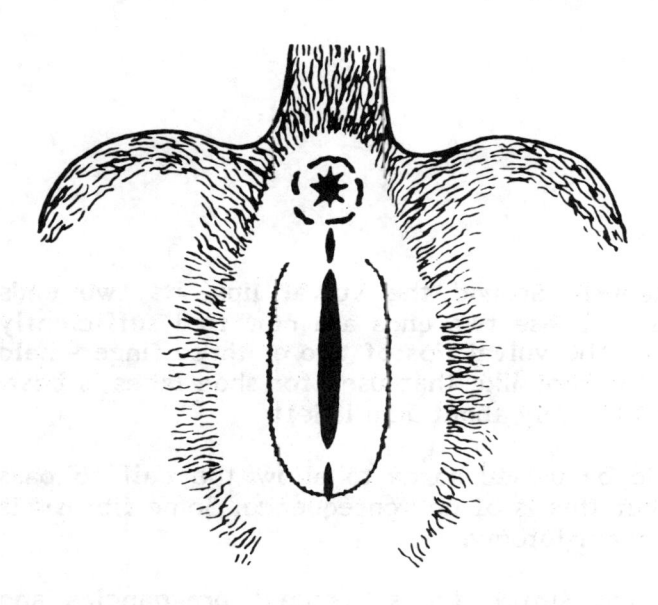

Fig. 7:8a

Whilst an assistant holds the tail aside two small (1cm long) incisions through the skin are made in the mid-line, one mid-way between anus and dorsal commisure, the other about 2cm below the ventral commissure.

Fig. 7:8b

The needle is introduced through the lower incision with its curvature in the lateral/medial plane and passed up through the deeper subcutaneous tissues on one side of the vulval lips so as to emerge at the upper incision. One hand should be placed in the vagina whilst this is done. The needle is threaded, withdrawn and unthreaded, thus laying a piece of tape subcutaneously.

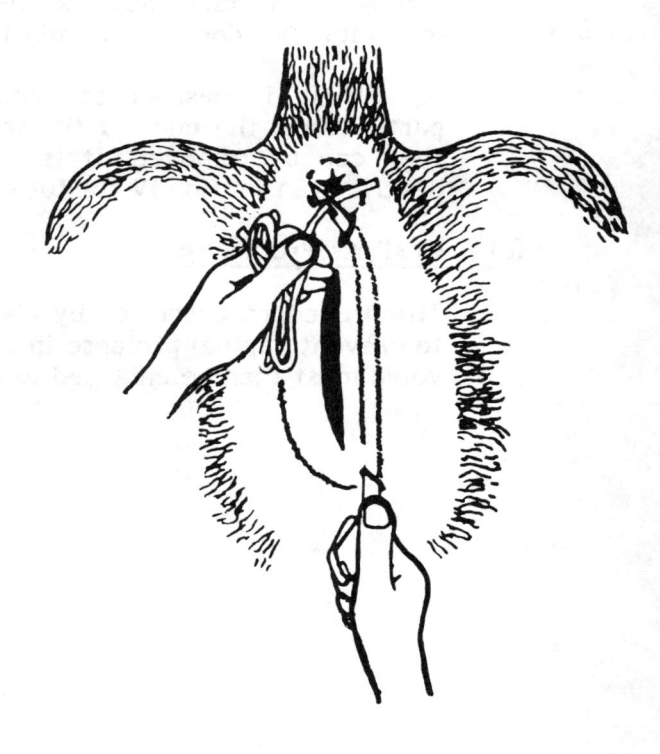

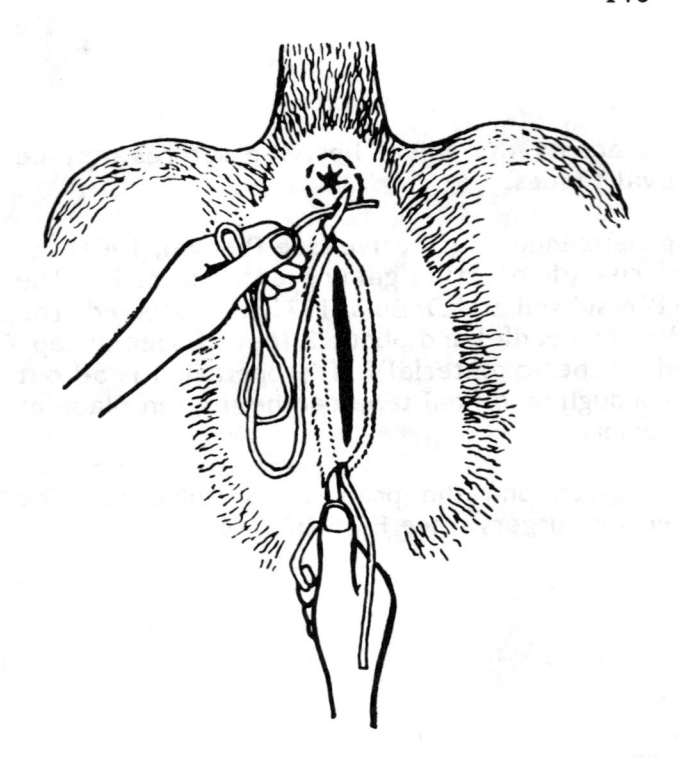

Fig. 7:8c

The procedures in Fig. 7:8b are repeated on the other side of the vulva.

The tape now lies subcutaneously around the vulval lips, its two ends protruding through the lower incision. These two ends are now tied sufficiently tightly to allow the insertion through the vulval lips of two or three fingers held together up to the second joint. A bow knot like that used for shoe laces is best. The free ends of the tape are cut so that only about 5cm is left.

At parturition the knot should be untied so as to allow the calf to pass through. The tape may disappear but this is of no consequence. Some fibrosis is inevitable but does not usually justify episiotomy.

If it is desired to replace the suture for subsequent pregnancies and parturitions, the ends of the tapes should be attached to suture material so that they can be removed. It is, however, preferable to use one of the techniques described in section IV for future pregnancies.

(e) Caslicks Procedure

The procedure described by Caslick for pneumovagina in mares can be employed to prevent vaginal prolapse in cattle. The technique is described on page 188. The vulva must then be enlarged to allow parturition and may be resutured afterwards.

IV PERMANENT RETENTION

These techniques are preferred if the cow is to be allowed subsequent pregnancies. They are best performed in the non-pregnant state but they may be undertaken in late pregnancy. Bearing in mind the hereditability of a tendency to prolapse it is doubtful whether the procedures should be encouraged.

(a) <u>Sub-mucous Resection</u>

This procedure was described by Farquarson (1949) and is applicable to chronic cases in which extensive necrosis and swelling of the vaginal mucosa are present. Under epidural anaesthesia, the damaged mucous membrane is resected, usually over a crescentic area and the incised edges coapted. Pierson (1961) comments that the procedure is time-consuming and is accompanied by much haemorrhage although Farquarson said that haemorrhage was minimal to moderate and was quickly controlled by suturing. Bouckaert and others (1956) state that in their hands the procedure was effective in only about 50% of cases. Pearson (1975) comments that, though the method is effective it occasionally results (perhaps due to faulty technique) in uncontrollable post operative straining and prolapse, not only of the sutured vaginam but also of the rectum.

(b) <u>Lateral Wall Fixation</u>

This technique was originally described by Minchev in Bulgaria but has been modified by a number of authors (see Pierson, 1961; Hentschl, 1961; Norton, 1969).

The objective of the procedure is to anchor the cranial vaginal wall of the sacro-sciatic ligament by ligatures passing from the vaginal lumen through the gluteal skin to the outside on each side. The ligatures, which may be of thick multisheathed nylon or of umbilical tape are prevented from pulling through the tissues by rolls of gauze impregnated with antiseptic (so Minchev) sterilized overcoat buttons (so Hentschl) or sterilized pads made of industrial belting (so Norton). It is important that the sciatic nerve is not damaged. The point of entry when pushing the needle through from outside is half-way along an imaginary line from the sacral tuber to the ischial tuber. (Compare Appendix III. Fig. 2).

Bouckaert and others (1956) report that the procedure is successful in 62% of cases.

The cow can, of course, calve and deliver fetal membranes past the devices. It is usually recommended that the sutures be removed after about 14 days, but which time it is usually stated that adhesions have developed which prevent recurrence of the prolapse. It is, however, also possible that the predisposing factors have been removed.

The procedure is not without its disadvantages. Firstly, there is as already mentioned the risk of damaging the sciatic nerve. Secondly, the method may not be adequate to retain the cranial portion of the vagina, thirdly, the cow may strain enough to pull the sutures through the mucous membrane and cause a fatal peritonitis and fourthly, a fatal peritonitis may develop as a result of sepsis.

(c) Fixation of the Vaginal Floor

In 1966, Winkler described a technique he had adopted successfully in 14 cows. It involved anchoring the vaginal floor and cervix to the pre-pubic tendon with non-absorbable suture material.

The author has not used Winkler's procedure but in fat non-pregnant Charolais cows which prolapsed the vagina he has anchored the vaginal wall to the 'ilio-psoas' muscles on the inner aspect of the shaft of the ilium with non-capillary, non-absorbable suture material through a left flank laparotomy incision. The suture was placed through the vaginal wall twice, once near the cervix and once about 5-8cm caudally so as to produce a tuck in the wall. In neither case was the needle allowed to penetrate the vaginal lumen so as to reduce the risk of peritonitis. In neither cow did the prolapse recurr.

IV CHOICE OF METHOD

With such a bewildering choice of methods available, it is difficult to know which to adopt in a given circumstance.

In ewes in late pregnancy, the truss methods are preferred although in a few difficult cases one of the intravaginal devices may be required. The simplest of cases can sometimes be controlled for a few days by tying the wool across the vulva. Ewes with prolapse are rarely seen at other stages of the reproductive cycle.

In cattle in the last few weeks of pregnancy may be satisfactorily treated by vulval sutures, but the method of choice for all pregnant cows must now be Buhner's perivulval sutures. This technique may also be used in cows with ovarian disease until that disease has been treated and cured. For non-pregnant cows and long-term retention, the author would prefer to anchor the vaginal wall at laparotomy.

D. REFERENCES

Arthur, G.H. (1975) Veterinary Reproduction and Obstetrics. 4th Edition. Bailliere and Tindall, p. 124.

Bassett, E.G. (1955) Pelvic dimensions of the Romney ewe. New Zealand Veterinary Journal 3 : 111-114.

Bassett, E.G. (1956) Vaginal Prolapse in the ewe. Proceedings of the 3rd International Congress on Reproduction Section 2 : 39-41

Bassett, E.G. and Philips, D.S.M. (1955) Changes in the pelvic region of the ewe during pregnancy and parturition. New Zealand Veterinary Journal 3 : 20-25.

Bassett, E.G. and Philips, D.S.M. (1955) Some observations on the pelvic anatomy of ewes with vaginal prolapse. New Zealand and Veterinary Journal 3 : 127-137.

Bassett, E.G. (1965) The anatomy of the pelvic and perineal regions of the ewe. Australian Journal of Zoology 13 : 201-241.

Bennetts, H.W. (1944) Two sheep problems on subterranean clover dominated pasture. 1 - Lambing trouble (dystocia) in merinos. 2 - Prolapse of the womb (inversion of the uterus). Journal of Agriculture of Western Australia 21 : 104-109.

Bierschwal, C.J. and DeBois, C.H.W. (1971) The Buhner method for control of chronic vaginal prolapse in the cow (Review and evaluation). Veterinary Medicine/Small Animal Clinician 66 : 230-236.

Bouckaert, J.H., Oyaert, W., Wijverkens, H. and Van Meirhaeghe, E. (1956) Prolapse vaginae bij het Rund. Vlaams diergeneeskunde Tijdschrift 25 : 119-132.

Bruere, A.N. (1955) Treatment of bearing trouble. New Zealand Veterinary Journal 4 : 170.

Edgar, D.G. (1952) Vaginal eversion in the pregnant ewe. Veterinary Record 64 : 852-858.

Fowler, N.G. and Evans, D.A. (1957) A new vaginal prolapse retainer for the ewe. Veterinary Record 69 : 501-502.

Farquharson, J. (1949) Vaginal Prolapse in the bovine. Proceedings of the 14th International Veterinary Congress, London 3 : 264-267.

Folley, S.J. and Malpress, F.H. (1944) Quoted by Edgar (1952).

Garm, O. (1949) Quoted by Edgar (1952).

Hentschl, A.F. (1961) The button technique for correction of prolapse of the vagina in cattle. Journal of the American Veterinary Medical Association 139 : 1319-1320.

Jones, B.V. (1958) Observations on the control of vaginal prolapse in a flock of breeding ewes. Veterinary Record 70 : 362-363.

Laing, A.D.M.G. (1945) 'Sleepy sickness' and bearing trouble in ewes - prevention by suitable methods of management. New Zealand Journal of Agriculture, 71 : 37-41.

McErlean, B.A. (1952) Vulvo vaginitis of swine. Veterinary Record 64 : 539-540.

McLean, J.W. (1956) Vaginal prolapse in sheep Part I. New Zealand Veterinary Journal 4 : 38-46.

McLean, J.W. (1957) Vaginal prolapse in ewes Part III. The effect of topography on incidence. New Zealand Veterinary Journal 5 : 93-97.

McLean, J.W. and Claxton, J.H. (1958) Vaginal prolapse in sheep Part IV. Cyclic changes in the vulva, vestibule and vagina during the year. New Zealand Veterinary Journal 6 : 133-137.

McLean, J.W. and Claxton, J.H. (1959) Vaginal prolapse in ewes Part V. Seasonal variation in incidence. New Zealand Veterinary Journal 7 : 134-136.

McLean, J.W. (1959) Vaginal prolapse in ewes Part VI. Mortality rate in ewes and lambs. New Zealand Veterinary Journal 7 : 137-139.

McLean, J.W. and Claxton, J.H. (1960) Vaginal prolapse in ewes Part VII. The measurement and effect of abdominal pressure. New Zealand Veterinary Journal 8 : 51-61.

Mayar, O.Y. (1958) A method of treating vaginal and uterine prolapse in the ewe. Veterinary Record 70 : 852.

Norton, E.S. (1969) External fixation of the bovine vagina after reduction of a prolapse. Journal of the American Veterinary Medical Association 154 : 1179-1181.

Pandit, R.K., Gupta, S.K. and Pattabiraman, S.P. (1983) Study on pelvic conformation in relation to prolapse of genitalia in buffaloes. Indian Veterinary Journal 60 : 463-466.

Pearson, H. (1975) The treatment of vaginal prolapse in the cow. Veterinary Annual 15: 54-56.

Pierson, R.E. (1961) Surgical procedures for correction of vaginal prolapse in cattle. Journal of the American Veterinary Medical Association 139 : 352-356.

Sobiraj, A., Busse, G., Gips, H. and Bostedt, H. (1986) Investigations into blood plasma profiles of electrolytes, 17B-oestradiol and progesterone in sheep suffering from vaginal eversion and prolapse ante-partum. British Veterinary Journal 142 : 218-223.

Walker, D.F. and Vaughan, J.T. (1980) Bovine and Equine Urogenital Surgery. Lea and Febiger. p.74.

Winkler, J.K. (1966) Repair of bovine vaginal prolapse by cervical fixation. Journal of the American Veterinary Medical Association 149 : 768-771.

Woodward, R.R. and Quesenberry, J.R. (1956) A study of vaginal and uterine prolapse in Hereford cattle. Journal of Animal Science 15 : 119-124.

Chapter Eight
THE CAESAREAN OPERATION

INTRODUCTION

The Caesarean Operation is probably the biggest piece of surgery which is carried out routinely in cattle and in any species constitutes a surgical interference of considerable magnitude.

The literature on caesarean operations is voluminous and no attempt will be made to review it systematically here. This chapter is, therefore, based primarily on the clinical experience of the author and his colleagues in the London and Liverpool Veterinary Schools.

One small linguistic point - the phrase 'caesarean section', is a tautology. Both the word 'caesar' and the word 'section' are derived from Latin verbs which mean to cut or to section. . . .

A. INDICATIONS

I   TO CURE A DYSTOCIA

A dystocia may be said to exist when spontaneous delivery is making inadequate progress. A caesarean operation may then be undertaken as an alternative to vaginal delivery because such would endanger fetus or dam.

Traditionally, factors causing dystocia are classified as fetal or maternal but in a proportion of cases both factors are present and may be inter-related. Thus a dystocia may be thought initially to be due to uterine torsion, but when the torsion has been corrected cervical non-dilation and a huge calf may be present.

(a)   Fetal Factors

Fetal oversize is the commonest indication for a bovine caesarean operation in general practice. Immaturity of the dam are often partly responsible, especially where the calf is male, but in some situations it is the 'double-muscled' calves of some breeds which are responsible for dystocia.

Malposture, malposition or malpresentation are not common indications for caesarean operations as most can be corrected and the fetus delivered normally. In the mare, many malpresentations can be dealt with satisfactorily with the mare under general anaesthesia and neither caesarean or fetotomy are required. However, transverse presentation of the foal, especially where the back is presented, is a positive indication for caesarean operation.

Monsters, especially the bovine Schistosomus reflexa, are an indication for caesarean operations. These are discussed again on page 160.

(b) <u>Maternal Factors</u>

The two most common maternal factors which are indications for caesarean operations are failure of the cervix to dilate and uterine torsion. These two may occur together, especially in cattle. Uterine torsion is discussed on page 160.

(c) <u>When</u>?

In order to achieve a good fetal and maternal recovery, caesarean operations need to be undertaken early in dystocias. The number of still-born young increases with the lapse of time between rupture of allanto-chorion and delivery and the maternal recovery rate is low if more than 24 hr have elapsed since 2nd stage labour began.

II TO PREVENT A DYSTOCIA

Caesarean operations can also be performed before the dam has entered labour so as to prevent a dystocia developing. Such operations are called 'elective'. The timing of the operation is critical if the young is to be saved. Ideally, the dam should be in first stage labour, easily recognised in cattle and sheep. In horses, the rapid strip test method devised by Cash, Ousey and Rossdale (1985) will be useful for predicting parturition. Fetal and maternal factors can be distinguished.

(a) Fetal Factors

Occasionally caesarean operation is indicated where it is known that a fetus is already oversized and dystocia will occur if parturition is allowed to proceed normally. The operation is best performed when the animal is in advanced first stage or is entering second stage labour, otherwise serious post-partum complications, especially metritis, ensue.

Cases of prolonged gestation in cattle and sheep are also indications for elective caesarean operations, but not in the mare, the majority of which will eventually foal live, normal term foals.

Hydrallantois is an indication for a caesarean operation. It is discussed further on page 162. Attempts to deliver mummified and macerated fetuses are less often made today. The former, however, remains a possible indication for caesarean in exceptionally valuable cattle where prostaglandins have failed (see page 149). Prevention of dystocia due to relative fetal oversize (and, therefore, of the need for caesarean) can be achieved by bull selection - in the U.K. A.I. bulls are evaluated for dystocia risk - and by heifer selection at the time of pregnancy diagnosis (see Deutscher, 1985).

(b) Maternal Factors

The commonest maternal factor which is an indication for an elective caesarean operation is pelvic deformity. Such deformity may be <u>congenital</u> (e.g. an extremely small birth canal) or <u>acquired</u> (e.g. bony deformity consequent upon pelvic fracture, rupture of the sacro-iliac ligaments, adhesions and scar tissue formation in the birth canal or excessive pelvic fat).

Uterine rupture recognisable prior to delivery of the young can also be an indication for caesarean operation as the laparotomy may allow the incision to be adequately repaired. (see also p.180 ff).

Mares with rupture of the prepubic tendon (p.64) and cows and ewes with ventral rupture may develop dystocia if allowed to proceed to normal parturition because their expulsive efforts are ineffective. Caesarean operation is then indicated, if economic or humane considerations so dictate.

III OTHER THAN FOR DYSTOCIA

Severe haemorrhage from the uterus or cervix or uterine rupture may be treated at laparotomy following a caesarean operation.

Preparturient paralysis or toxaemia (whether or metabolic or infectious origin) are occasionally indications for caesarean operation if the expectation is that the dam will, thereby, recover. However, slaughter of the dam and recovery of the live young at slaughter may be a better economic proposition.

Caesarean operations are also performed to obtain Minimal Disease and Specific Pathogen Free pigs for commercial purposes and to obtain gnotobiotic animals for research purposes.

B. ALTERNATIVES

Caesarean operation is by no means the only way of dealing with a dystocia in which vaginal delivery is making no progress. This section considers briefly some of the alternatives and the factors that may make them preferable to a caesarean operation.

I FETOTOMY

Fetotomy does not appear to be as popular in the British Isles as on the continent of Europe, although how much this is a reflection of agriculture or of veterinary school attitudes is difficult to say.

Fetotomy is definitely indicated in cows and horses in which the fetus is dead and a single malposture, such as uncorrectable lateral deviation of the head, is the sole cause of dystocia. Vandeplassche, Bouters, Spincemaille and Bouters (1978) comment that they only perform one caesarean operation for every 15 fetotomies. Cases of hydrocephalus in piglets (in which the head can usually be crushed or broken) or of rotten fetuses in sheep (which can usually be easily dismembered) are not cases for Caesarean operation because vaginal delivery is easy.

Anasarcous calves or cases of dropsical fetuses are in an intermediate group. Where, by a single or by a few incisions, the fluid can be released, then vaginal delivery is usually not too difficult. In cases in which the extent of the fluid is too great for the fluid to be readily drained off, then caesarean operation may be the better option if the dam is to be saved.

Emphysematous fetuses also fall into an intermediate class. Some such cases can be delivered by fetotomy but, in others, peri-vaginal swelling, cervical non-dilation or simply gross fetal oversize restrict the relative size of the birth canal so that fetotomy cannot be carried out. Emphysematous calves are discussed again on page 161.

Probably the most contentious area, however, is whether fetotomy or a caesarean operation is the better way to deal with a dead young. One approach says that a dead fetus is a positive contraindication for a caesarean operation as it avoids any peritoneal contamination and uterine scarring. Others would hold that for those skilled in caesarean operations, the caesarean offers the dam a better prognosis especially as one knows within limits how long a caesarean is going to take whereas one cannot always be sure that a fetotomy will prove simple. Truly comparable figures of maternal survival and future fertility are difficult to obtain because criteria are not adequately defined and results rarely evaluated statistically. To paraphrase Jane Austen, each school of thought probably begins with a little bias towards its own point of view, and upon that bias builds every circumstance in favour of it which has occurred within its own circle.

Techniques for fetotomy are excellently described by Bierschwal and DeBois.

II   PELVIC SYMPHYSIOTOMY

Splitting the pelvis of the immature heifer (less than 26 months of age) with a bone chisel in order to deliver an oversize calf has been described (amongst others) by Harsch and Hanks (1971), and is apparently widely practised in the range country of North America.

Although the authors quoted state that their heifers were merely confined with their calves for several days, it is clear from personal conversations that many animals stay recumbent for many days.

It seems unlikely that such an approach would be acceptable in the British Isles with its more humanitarian attitude to animals.

III   ANTIBIOTICS AND DELAYED OPERATION

Ford (1965) has described treating cows with dystocia for several days with systematic antibiotic and then operating. No mention was made of whether the cows were straining at the time treatment was initiated.

In the U.K. such an approach would probably and justifiably result in a successful prosecution for cruelty.

IV   CERVICAL SECTION

Pearson (1971) has described cutting the cervix of some cows in dystocia in which cervical non-dilation was the only impediment to vaginal delivery.

This procedure is only indicated if:-

(i)   the birth canal caudal to the cervix is sufficiently dilated to allow delivery.

(ii) the cervical rim is thin and stretches like a sleeve on the fetus when traction is applied.

(iii) the fetus is not excessively large.

Should the cervix be thick or the fetus excessively large, cervical section is contra-indicated as the risk of uncontrolled tearing is great.

The technique is to stretch the cervix by traction on the fetus and then to incise the full depth of the rim at one point only. Pearson still reports (personal communication) that the technique is a valuable alternative to caesarean in selected cases of non-dilation of the cervix.

V MEDICAL THERAPY

Oxytocin injections, repeated if necessary at appropriate intervals, are the usual method of dealing with dystocias in the sow, when the immediate obstruction to delivery has been relieved by vaginal manipulation.

Delivery can be induced in the mare by intramuscular oxytocin and this may occasionally be useful.

'Spasmolytics' such as promaquezine fumarate ('Myspamol', May and Baker), hyoscine-n-butylbromide (in 'Buscopan Compositum', Crown Chemicals) and methindizate (in 'Isaverin', Bayer) have been rather vaguely recommended in the literature for treating cervical non-dilation in cattle and sheep but there is little clinical evidence that they are effective in more than a small proportion of cases.

Corticosteroids and Prostaglandins have revolutionised our ability to control the timing of parturition and in some situations they offer a reasonable and practical alternative to caesarean operation.

(i) Corticosteroids require a live fetus and so are effective in cases of bovine hydrallantois (see Vandeplassche, Bouters, Spincemaille and Bonte, 1974) and other cases in which immediate delivery is not required.

(ii) Prostaglandins may be effective in inducing oestrus in cattle with mummified or macerated fetuses if a corpus luteum is present and so opening the cervix/the fetus may then be delivered spontaneously or by traction. They may also be useful to induce premature delivery of a potentially oversize calf.

(iii) The administration of oestrogens has gone out of favour as a means of inducing parturition but it is still a possibility that might be considered in a cow with a mummified fetus in which prostaglandins have failed.

VI SLAUGHTER

It may be that the cost of caesarean operation or any of the other procedures listed above will cost the owner more than they can afford or are willing to pay, or that the risks of attempting delivery (e.g. with emphysematous calves with uterine rupture) are so great as not to be economically justified. In such cases, slaughter will be indicated to salvage as much as possible, though a parturient animal may set badly.

VII EPISIOTOMY

Episiotomy is described and discussed in chapter 10 (p.193).

VIII GENERAL REMARKS

It is hoped that the preceding remarks have indicated how difficult it can be to make hard and fast rules about when a caesarean operation is indicated and when it is contraindicated. It is important that each case be assessed on its own particular features, bearing in mind not only the dystocia present but also the state of health of the dam, the viability and the value of the young, the economic circumstances in which the case is met, the attitudes of owner and veterinarian and the skills of the latter.

Moreover, the operation cannot be considered in isolation from the future potential of the animal. Section D of this chapter details the post-operative course and complications of caesarean operations and these must also be borne in mind when making the decision as to whether to operate or not.

Nevertheless, it is important to make the appropriate decision quickly. Unsuccessful attempts at alternatives to caesarean operation can jeopardize the chances of a successful caesarean. It can rarely be absolutely wrong to do a caesarean operation, but it can often be wrong not to do one.

C. TECHNIQUE

I ANAESTHESIA AND RESTRAINT

(a) Cow

A place should be chosen for the operation which is clean and in which there is adequate light. The dirty loose box with a single low wattage bulb is totally inadequate. It is now possible for a few pounds to invest in a powerful, mains-rechargeable, battery operated torch and these provide an excellent source of light. The stockman should be discouraged from sweeping down cobwebs or spreading straw about gaily as both contribute considerable airborne contamination for several hours.

The author's preference is to perform the operation with the animal standing and one animal attendant at the animal's head. A rope is placed round the right hind leg and laid underneath the animal's body (Fig. 8:1) so as to be able to pull the animal into right lateral recumbency should she decide to go down during the course of the operation, approximately 5-10% doing so.

Other methods are possible and the eventual choice in any given situation may depend upon the personal preference of the operator, his height and the assistance available. Some advocate right lateral recumbency with the head held down, the two fore-legs and right hind leg hobbled together about the metacarpus and metatarsus, and the left hind leg held backwards by a second rope and hobble. Others prefer to have the animal sedated in sternal recumbency but tilted slightly to the right and with the left hind leg pulled backwards. (This is the author's preference for the recumbent cow which will sit quietly). Some of these possibilities are illustrated in Noordsy (1979).

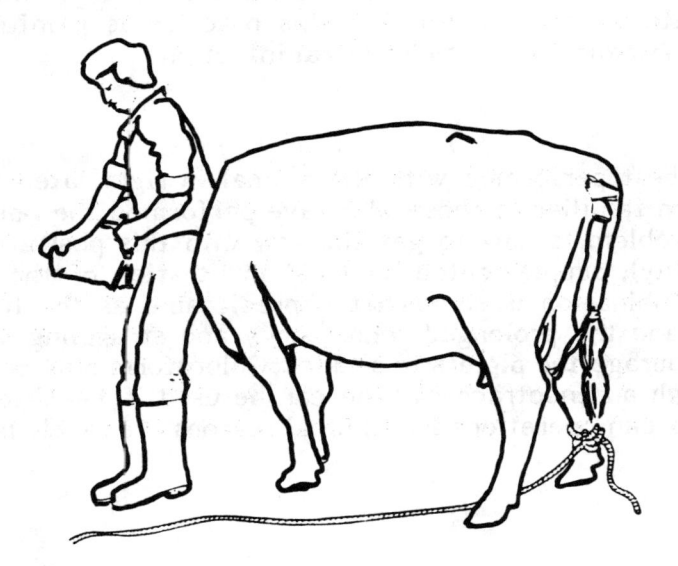

Fig. 8:1 Showing cow restrained for caesarean operation in the standing position. The tail is restrained by a bandage. A rope, tied round the right hind leg lies on the floor below the cow and is placed so as to be able to pull the cow into right lateral recumbency should she go down. An assistant restrains the head.

For the author's preferred site (see p.154) the left flank of the animal is clipped from the level of the base of the last rib cranially, to the tuber coxae caudally and from the middle of the back to below the fold of the flank.

If the cow is straining, low epidural anaesthesia should be given. If not, the tail should be tied to a leg to stop it flicking the wound. Either xylazine (at a does of not more than 25mg for a 500 kg cow) or chloral hydrate by mouth (30-60 gm for a 500 kg cow) may be used for sedation. Xylazine has the disadvantage that it can increase uterine tone even if spasmolytics are given simultaneously so making manipulation more difficult, although this difficulty is said to be overcome by doubling the dose of relaxant to 460mg (Ahlers, Luhmann and Andresen, 1971). Paravertebral anaesthesia of the 13th thoracic and first and second lumbar nerves is the method of choice (Appendix IV). Although field infiltration or 'L-blocks' are used they do not achieve as good analgesia as paravertebral anaesthesia.

See under (d) Mare for remarks on general anaesthesia.

(b) Ewe

The operation is best performed with the sheep in right lateral recumbency on a bale of straw or a table. In the latter case, the legs may be tied down (the hind legs extended slightly backward) and the head either held by an animal attendant or secured by a rope laid over the neck just behind the jaws and tied to the table.

Sedation is rarely required and anaesthesia may be by paravertebral or by field infiltration. Epidural anaesthesia is rarely necessary.

The wool is clipped over an area similar to that for cattle. With care the clipped wool can be rolled forwards over the chest and unrolled again after surgery and tied back into place - this provides adequate post-operative protection against the cold. Some practitioners prefer to pluck the wool only over the immediate site of the incision but this practice is painful and results in bruising and the subsequent risk of clostridial infection.

(c) Sow

The operation is best performed with the animal in right lateral recumbency as the position is then familiar to those who have performed the operation in cattle and sheep. The problem is how to get the sow into this position! Azaperone (1 mg/kg intravenously), supplemented by local infiltration of the site of incision works but the combination is somewhat unpredictable as the higher does rates must be avoided and the prolonged recumbency and struggling which commonly follow do not encourage the piglets to suckle. Chloroform may be used or gaseous anaesthesia through an endotracheal tube can be used if facilities are available. Fortunately, caesarean operations for clinical reasons are rarely performed in this species.

(d) Mare

General anaesthesia is to be preferred. (The author's experience makes it impossible for him to agree with most Continental veterinarians (see Vandeplassche, 1973) that general anaesthesia is detrimental). Xylazine is to be preferred to a promazine derivative or a butyrophenone for premedication as the blood pressure is not depressed. Both thiopentone (see Copland, 1976) and chloral hydrate depress respiration of the fetus and, therefore, the author's choice with a live foal is methohexitone for induction followed by light gaseous anaesthesia. Copland (personal communication) has shown that the choice of induction agent has a profound influence on the liveliness of the young. Although live foals may be born following induction with thiopentone or chloral, such foals can prove difficult to rear whereas those from mares induced with methohexitone are almost as bright and alert as foals born naturally. These remarks apply equally to cattle and sheep (see, for example, Tavernor, Trexler and others, 1971).

For the site of operation, see p.155.

II THERAPY

(a) Antibiotics

Asepsis is unachievable in veterinary practice - even when operating upon sows to procure gnotobiote piglets at a level of cleanliness far above that achieved in most clinical situations, swabs of the abdominal incision were sometimes found not to be sterile (Trexler, personal communication) - and in most clinical cases manual interference by stockman and veterinarian has contaminated the uterus.

The author's practice, therefore, is to administer systematic antibiotics before surgery commences. There is little advantage in cattle or pigs in using anything other than a streptomycin/penicillin mixture, in sheep in using other than long acting penicillin and in horses in using other than crystalline penicillin, the last mentioned being given intravenously. In addition antibiotic pessaries are placed in the uterine horn(s) before closing the uterine incision. The use of antibiotic pessaries is particularly important in elective caesareans in which the cervix has not yet opened. 6mg of crystalline penicillin in solution is placed in the peritoneal cavity prior to closure of the abdomen. Local application of antibiotic powders to incisions before suturing is contraindicated.

(b) Smooth Muscle Relaxants

Isoxsuprine ('Duphaspasmin', Duphar Veterinary Limited) is being increasingly used in cattle to aid relaxation of and, therefore, exteriorisation of the uterus and its introduction has been hailed as an advance comparable to the advent of sulphonamides (Aehnelt, Grunert and Andresen, 1971). However, there is an increased risk of haemorrhage from the flaccid uterus. Moreover, the basic principle underlying its use - that it is necessary to exteriorize the uterus before incising it - does not apply to the special case of an emphysematous fetus. In addition, the sceptic can point to the fact that no adequate double-blind trial has been used to assay its true efficacy.

III OPERATIVE TECHNIQUE

(a) Choice of Operation Site:

Historically the mid line or paramedian approach was the first to be widely used for cattle and sheep but when paravertebral anaesthesia became legal in the U.K. (1954) the flank site became preferred. Early experience showed that the logically better right flank approach (the pregnant uterus is displaced to the right by the rumen) usually meant getting tangled up with small intestine and nowadays the left flank is almost universally used.

A number of different approaches through the left flank have been described. The classical one is a vertical one starting 7-10 cm below the lumbar processes and approximately in the middle of the sub-lumbar fussa. For many years now the veterinary school in Liverpool has preferred an incision at an angle of about 60° to the spine starting about 10 cm away from the tuber coxae on such a line (Fig. 8:2).

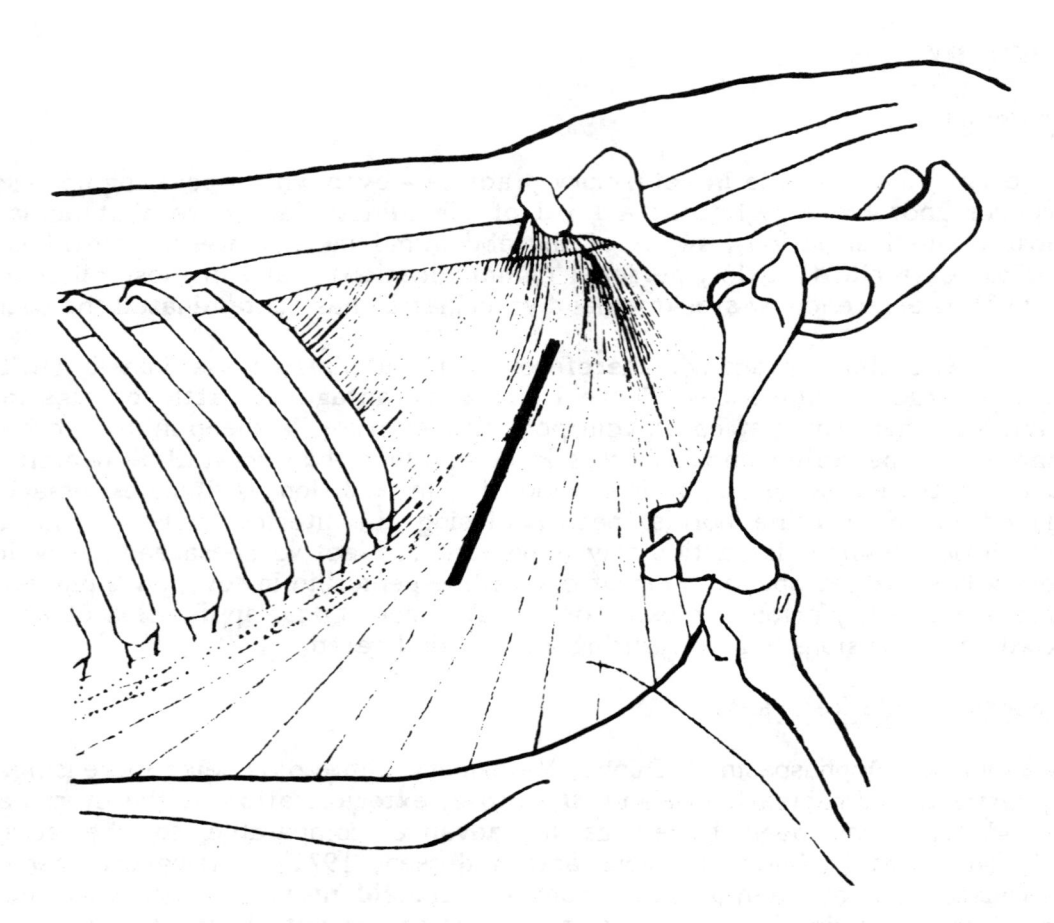

Fig. 8:2 Lateral view of internal oblique muscle of cow showing preferred line of incision for caesarean operation.

This approach has the particular merits that the internal abdominal oblique muscle can be split along its fibres and there is, therefore, less haemorrhage and that a retracting uterus can be followed back into the pelvis more easily. The disadvantages of this approach over the conventional vertical one are that if the incision is extended too far caudo-dorsally the circumflex iliac artery may be damaged and that if the incision is extended too far cranio-ventrally anaesthesia may be inadequate. In practice, however, these disadvantages are of little significance. The incision must be of adequate length - 30-40 cm in the cow, 15 cm in the sheep.

Low flank incisions (lateral to the mammary vein but below the fold of the flank) or horizontally just above the fold of the flank have been described as being of particular use for delivering emphysematous calves as they allow easier exteriorisation of the uterus. Noordsy (1979) illustrates some of these approaches. However, these laparotomy incisions are difficult to repair, healing may be poor and other techniques are available for emphysematous calves. (see page 161).

The caecum of the horse lies on the right and a left-flank approach avoids it. The small sub-lumbar fossa of the horse, however, makes flank operations difficult. The incision should run at least 10-15 cm away from the last rib and approximately parallel to it. The mid-line is, in fact, a far easier surgical approach and can be adequately repaired in even the largest horses. The incision is made for 30-40 cm cranial to the mammary gland.

There seems to be no anatomical preference of sides in the pig although the left flank is generally preferred because of its familiarity. The mid-line is hopeless because of the mammary glands. The incision should be made vertically or on a slope as in the ruminant.

(b) The Laparotomy Incision

The layers should be worked through one at a time controlling haemorrhage by haemostats and/or ligatures as necessary.

The layers in the flank are:

(i) skin;

(ii) cutaneous muscle (which fans out from the stifle and is, therefore, thickest in the ventral part of the incision);

(iii) external abdominal oblique muscle (almost uniformly thick muscle);

(iv) internal abdominal oblique muscle (very thick muscle dorsally, tendinous ventrally) - with the sloping approach in ruminants this muscle may be split along its fibres;

(v) transverse abdominal muscle (thinnish muscle tending to become tendinous ventrally) and

(vi) peritoneum (substantial in the cow) - care should always be taken incising peritoneum to reduce the risk of incising viscera.

Variable amounts of fat are present - the dairy heifer usually has little; the multi-parous beef cow inches of it, particularly below the skin and retroperitoneal.

The layers in the mid-line are skin, subcutaneous fat (often crossed by large blood vessels), linea alba, retroperitoneal fat and peritoneum (delicate in the mare). Care is needed with anastomosing branches of the mammary veins.

(c) Incising the Uterus

This part of the procedure gives most trouble to novices on 'simple' cases and to experienced operators in 'difficult' cases. Once the peritoneal cavity has been entered the uterus should be located and in the ruminant and horse the pregnant horn identified, and assessment is made of the possibility of bringing part of the pregnant uterus to the outside world.

In the ruminant the uterus may be located by passing a hand through the dorsal part of the abdominal opening towards the pelvis. The pregnant uterus usually lies caudomedially to the rumen and below the level of the pelvis (Fig. 8:3).

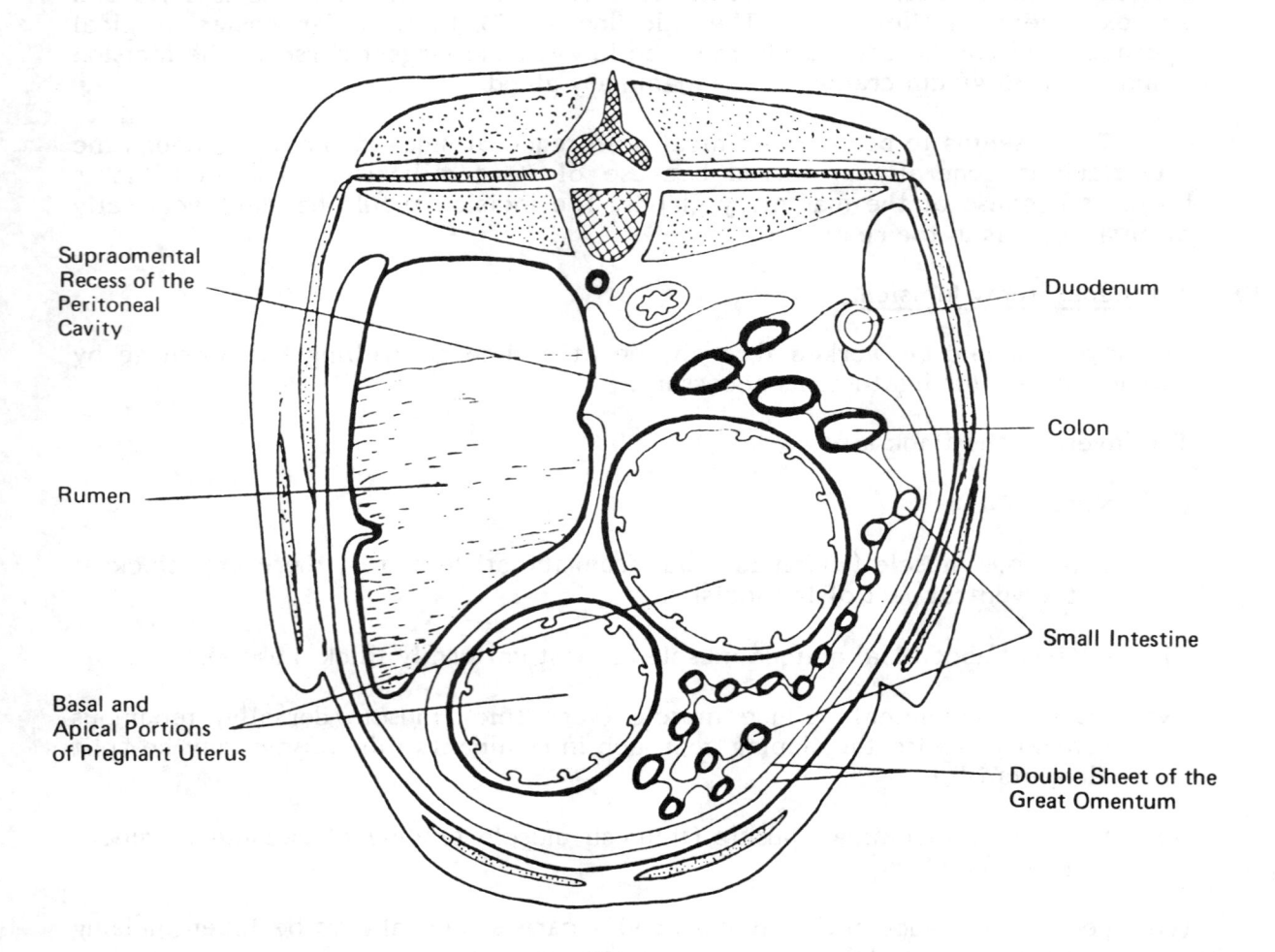

Fig. 8:3 Diagrammatic transverse section through the abdomen at the level of the 6th lumbar vertebra of a cow in the last month of pregnancy. (After Rasbech, 1957)

The uterus itself is then identified between the broad ligaments and its greater curvature identified. A note is made of which horn contains the pregnancy - the right-horn pregnancy is often more difficult to exteriorize than the left. Many authorities talk glibly about exteriorizing a portion of the bovine uterus before incising it but there can be few people who can lift 100-200kg of uterus and contents. Experience confirms Messervey and other's observation (1956) that if exteriorisation is difficult, resulting in undue handling of the uterus, it is preferable to incise the uterus in the abdominal cavity. The most convenient way of attempting exteriorisation of the uterus in the horse and ruminant is to grasp a fetal limb in the ovarian end of the uterine horn through the uterine wall and apply gentle traction to it. (The uterine incision must not be made over the cervical end of the uterus as there is an increased risk of uncontrollable tearing and such an incision is difficult to repair).

In the horse and sheep exteriorisation of part of the uterine horn containing a fetal limb is relatively easy.

In cattle, only in elective ceasareans (or perhaps those given isaverine) and in cases in early second stage labour, will it be possible to bring a fetal appendage near to the abdominal incision - in delayed cases movement of the uterus is often almost impossible.

In the pig, great care must be taken to avoid bringing too much uterus to the outside - the inexperienced operator is likely to be over enthusiastic! The author prefers in this species to identify the middle two piglets in one horn and exteriorise these two only.

The incision in the exteriorised uterus is most conveniently made with scissors although a scalpel may be used in the pig. When the uterus cannot be exteriorised (the majority of bovine cases) a Robert's knife is useful for making a blind incision. To avoid losing the knife a loop of suture material may be attached to it and wrapped round the wrist. In the ruminant and horse a fetal toe is identified and the beak of the knife pushed through the uterine wall above it and away from the ovary, ensuring that the full thickness of the uterus has been penetrated. The knife is then advanced along a limb up to at least several cm beyond carpus or tarsus. In the sow, the incision should be made between conceptuses. It is particularly important that the incision be made at least as long as indicated to avoid tearing the uterus as the young is removed from it.

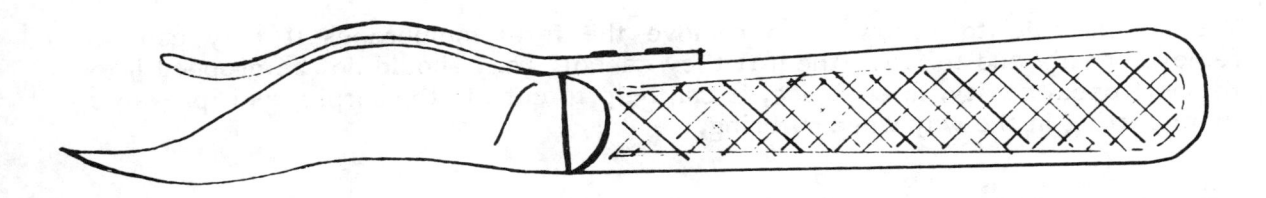

Fig. 8:4 Roberts embryotomy knife — useful for incising the bovine uterus when it cannot be exteriorised.

The incision itself should be made, if possible, so as to avoid cotyledons in the ruminant (they bleed profusely) - they can be identified by touch, although it is difficult to do this when making a blind incision in the cow. The incision should also avoid going too near the ovary as the risk of adhesions causing later infertility is greater. In the pig and the horse the incision should be made along the antimesometrial border - this can be identified visually. In all species it should be made along the long axis of the uterus. It should not be forgotten that besides incising uterine wall, the allanto-chorion and ammion must also be cut.

(d) Removal of the fetus(es)

The operator should try to get both hind feet or both fore feet and head extended and through the incision before applying traction. A surgical assistant is required for this part of the procedure. The author prefers not to use ropes or chains but to have traction applied directly to the fetus. The fetus should be delivered gently, avoiding severe traction and enlarging the incisions, particularly that in the skin, if necessary. The operator himself should be ensuring that skin and uterine incisions are of adequate size and should be holding onto the uterus so that it does not disappear back into the abdominal cavity after extraction of the young. The uterus is then searched for a second or third young.

In the horse, the foal should be kept attached through its umbilical cord after removal until an arterial pulse is no longer palpable in the cord to allow some of the placental pool of blood to be transfused into the foal (some practitioners do this with cattle also). The umbilical cord may be tied with tape rather than suture material as the tissue is very friable but in cattle and sheep the short umbilical cord usually snaps easily and safely. Should bleeding occur, it may be necessary to apply haemostats to the paired arteries and single vein.

Practical considerations in the pig suggest that the incision is best made in the middle of one uterine horn, all piglets removed from that horn, the incision sutured, and the procedure repeated with the other horn. Ony when it is believed to be empty should a complete horn be exteriorised to confirm that it is so. Piglets can be extracted most simply by putting hand and forearm through the uterine incision into the uterine lumen and reaching along the horn - both hind legs or the head are grasped. Milking the piglets along a horn from outside bruises the uterine wall and is, therefore, contraindicated.

(e) Fetal membranes

The simple rule to follow is to remove the fetal membranes if they can be removed easily and to leave them if they cannot. They should not be dropped into the peritoneal cavity. In any case, it can help to cut off the surplus as it prevents them getting in the way when suturing.

(f) Suturing the uterus

Before suturing is undertaken, antibiotic pesseries should be placed in the uterine lumen.

Any major bleeding points in the uterine incision should now be indentified and dealt with. In the mare, the endometrium is only loosely attached to the myometrium and the two are separated by a large plexus of veins. These may bleed profusely and cannot be controlled by conventional clamping and ligation.

After Vandeplassche and his colleagues (1962) had had several mares die of haemorrhage from these veins after caesarean they devised a procedure to prevent such a catastrophe. The allantochorion and endometrium are carefully separated for about 2cm all round the edge of the incision. A simple over and over pattern penetrating all three layers of the uterine wall is then placed all round the edge of the incision. However, in the author's experience, most of the oozing has stopped by the time the suture is completed and he no longer uses this procedure, preferring an oxytocin infusion immediately after suturing the uterine incision (see page 160) to achieve rapid uterine involution.

A continuous inverting suture such as a Lembert (in which the needle passes at right angles to the incision) or a Cushing (in which the needle passes parallel to the incision) is used. In the ruminant the suture should begin at the cervical end of the incision as this can become readily retracted into the abdomen. In all species the suture line should begin and end beyond the limits of the incision to ensure adequate inversion. One layer is usually enough but if uterine tone is poor, interrupted sutures and/or a second layer may give better security. Absorbable suture material should be used - De Bois at Utrecht has argued that plain gut is preferable to chromic as the latter lasts a long time. In the mare, in particular, it is important that the sutures are not taken so deep as to anchor fetal membranes; the relatively thich uterine wall of the cow makes this unlikely. In all species it is important that fetal membranes should not be trapped between two edges of the incision.

Once suturing is completed , the uterus is cleaned by swabbing and returned to the abdomen. Oxytocin may now be given as described below (page 160).

(g) Repairing the abdominal wall

Antibiotic in a soluble form (not as an ointment) is first placed into the peritoneal cavity.

Many techniques are possible, but the following gives excellent results in the cow.

(i) The peritoneum and transverse muscle and tendon are sutured together with a continuous suture of 7 metric chromic gut. (Sloss and Dufty, 1978, have shown that adequate suturing of peritoneum is the most important single factor in the prevention of sub-cutaneous emphysema).

(ii) The internal oblique muscle is similarly sutured and the external oblique and cutaneous muscle are then also sutured together with a continuous suture.

(iii) The skin is sutured with horizontal mattress sutures with a suitable non-absorbable material. Frerking, Andresen and Geyer (1967) have presented evidence that skin clips are preferable to sutures in that, as the skin is not penetrated, there is less post-operative sepsis but sepsis from skin sutures has not been a problem for this author.

Thus, care is taken to close dead space within the flank. The method has the advantage over the commonly used 'vertical mattress' that the tension is taken up layer by layer and there is minimal post-operative swelling. It takes little if any, longer than the vertical mattress technique.

Similar techniques can be used in the pig and sheep.

The mare poses several problems if the flank approach is used. Firstly, post-operative swelling is normal and is exaggerated by using cat gut and secondly, synthetic absorbable suture materials are not available in thick enough gauges to avoid their Gigli-wire like effect. To those accustomed to handling stainless steel, this may be the best material. The author has had satisfactory results using a Teflon-coated braided polyester prepared for closing the human sternum (gauge 7 metric) ('Ethibond', Ethicon Ltd.). This is used for closing the linea alba, and synthetic absorbable suture material is then used in the subcutaneous tissue and the skin. It is not necessary to suture peritoneum independently or at all (Swanwick and Milne, 1973).

(h) Ecbolics

Oxytocin alone or in combination with ergometrine (the latter is no longer available in the U.K. without oxytocin) after suturing the uterus can help to reduce uterine size and may stimulate expulsion of the fetal membranes. The dose rate is 10 i.u./50 kg and may be given by intramuscular injection. In mares the preferred route is by intravenous drip, (see Vandeplassche, Spincemaille and Bouters, 1971), the calculated dose being given in 500-100 ml of saline whilst the laparotomy wound is repaired. In cases in which the cervix is closed, attempts should be made to penetrate the cervix and rupture the allontochrion - this is not usually difficult (see also p. 165).

IV SPECIAL CASES

(a) Monsters

The commonest monster of all is the Schistosoma reflexus seen in cattle and occasionally in other ruminants. The majority of these calves are smaller than usual but the ankylosis they have presents problems to the operator. Larger skin and uterine incisions than normal are necessary, and partial fetotomy with wire before delivery may be necessary in some cases. These remarks apply also to other monsters.

The decision to perform a Caesarean on an unidentified monster may be a very difficult one to make as it may be discovered at operation that the fetus cannot be delivered except with so great a trauma to the dam that her recovery is unlikely.

(b) Irreducible Uterine Torsion

Occasionally, the techniques of rolling, abdominal ballotment and rotation of the fetus per vaginam fail to correct uterine torsion in the cow - such cases are most often those in which there has been delay in seeking veterinary advice (see Pearson, 1971, for a review of uterine torsion). One solution in such cases is to do a laparotomy as for a caesarean operation, and then attempt to correct the torsion within the abdomen.

Arthur (1975) has described his approach as follows:-

'For a twist to the left (i.e. anticlockwise), the hand is passed down between the uterus and the left flank and a fetal hand-hold is sought whereby an attempt is made first to 'rock' the uterus and then to rotate it by strongly lifting and pushing to the right. For a twist to the right (i.e. clockwise) the hand is passed over and down between the uterus and the right flank and as before a swinging manoeuvre is followed by pulling upwards and to the left'.

Arthur further comments that 'owing to the oedema of its walls the uterus is unusually friable'.

If the torsion is corrected, vaginal delivery should then be attempted.

If the vaginal delivery is not possible (perhaps because of cervical non-dilation which is not amenable to cervical section - see page 148 - or the torsion is irreducible) then a Caesarean operation should be performed. It will be possible to exteriorise such a uterus before incising it so the incision must often be made blind. An occasional sequel to this is that it is then difficult to suture the uterus as the incision has become inaccessible, the torsion having corrected itself spontaneously as the calf was delivered. Even if the torsion did not correct itself, it can still be difficult to exteriorise the uterus for suturing. Uterine inertia is often present and tension on the sutures may result in them pulling out.

In some cases, laparotomy will reveal that the uterus is so congested or so friable that surgery is unlikely to save the dam -immediate slaughter is then indicated.

Uterine torsion is occasionally encountered in sheep.

In the mare, uterine torsion may occur weeks or months before foaling as often as it occurs near to foaling (Spincemaille, Vandeplassche and Bouters, 1970). Rolling, even under general anaesthesia is contra-indicated because the risk of uterine rupture is very high. Repositioning at laparatomy and allowing pregnancy to continue is indicated in cases seen prior to full term provided that the uterus and foal are healthy. The technique can be performed in standing mares under sedation and local anaesthesia through a small (15 cm) flank laparotomy incision. By inserting the whole of a forearm underneath the uterus, the uterus can be readily untwisted (Vandeplassche, Paredis and Bouters, 1961). In cases at term, caesarean operation is indicated in preference to rolling. The condition is much more likely to be fatal in horses than in cattle, primarily because uterine rupture is much more likely and more animals will die of shock.

(c) The Emphysematous Fetus

Much has been written in the continental literature on techniques for dealing with the emphysematous fetus in which fetotomy is impossible.

In such cases the cow may be quite healthy, even though there are considerable quantities of toxins and bacteria in the uterine lumen. These are apparently reaching the systematic circulation so slowly that the reticulo-endothelial system is coping with them. If, however, the toxins are allowed to spill into the peritoneal cavity, they are absorbed rapidly, the liver cannot cope and the cow quickly becomes toxaemic. Moreover, the organisms present result in a lingering peritonitis (see also p.166ff).

One approach to such cases is, therefore, to attempt exteriorisation of the uterus at all costs and to this end a low flank incision is used (see page 151 and Oehme, 1967). The approach outlined below, however, gives the approach of the Liverpool School (as developed by Wyn-Jones) to such a case.

The conventional sloping incision is made in the flank. If bits of fetus or fetal membranes are found in the peritoneal cavity, then the operation should be abandoned as the uterus has ruptured. The presence of large amounts of peritoneal fluid, perhaps haemorrhagic, or of fibrin flakes is of no immediate concern. An incision of adequate length is made in the uterus over a suitable fetal appendage. The uterus is brought as near to the exterior as possible and the calf extracted, taking care to reduce to a minimum loss of hair, hoof and fetal membrane into the peritoneal cavity. Following delivery of the calf, the uterus is sutured in the normal way with antibiotic oblets placed inside. Three litres of warm 0.9% saline are emptied into the peritoneal cavity and then baled out. This lavage procedure is repeated at least once. Finally, twice the normal therapeutic dose of crystalline penicillin in solution is placed in the peritoneal cavity.

Careful trimming of the contaminated edges of the muscle incisions may be practised in severe cases before closure of the abdominal wall as described previously.

Post-operative care (see Section D) needs to be intense and carefully planned, (e.g. a second intraperitoneal injection of crystalline penicillin should be given on the second post-operative day), but the majority of cases make a satisfactory recovery. Post-operative diarrhoea is not unusual, presumably indicating peritoneal irritation, but usually subsides within a week.

(d) Mummified fetus

In these cases, the major difficulty is that the uterus is small and impossible to exteriorise. All incising, extracting and suturing may, therefore, have to be carried out intra-abdominally.

(e) Hydrops

This rare condition occurs in both cattle and horses.

(i) The cow

Hydrallantois in the cow develops within the last 3 months of pregnancy and is characterised by excess accumulation of fluid in the allantoic cavity. The aetiology and pathogenesis of this condition have been discussed by Arthur (1957), Skysgaard (1965) and Vandeplassche and others (1965).

The accumulation of fluid progressively interferes with respiration and appetite so the cow becomes progressively weaker and emaciated and, if treatment is delayed, may become recumbent and die.

Diagnosis is usually easy because the excessive fluid in the abdomen can be readily appreciated and rectal examination reveals a grossly distended uterus, perhaps extending back into the pelvic canal.

Whether treatment is attempted or not depends upon the value of the carcass, the potential milk yield and the state of the cow. Recumbent cases should be slaughtered. Cases with marked distension should be treated surgically so as to obtain rapid resolution and less severe cases may be treated medically.

Vandeplassche and others (1974) successfully treated 17 of 20 cases by giving an intramuscular injection of corticosteroid (see page 149). After 4-5 days the cervix relaxed and the cow was given oxytocin by intravenous drip (40 i.u. in 1 litre of saline over 30 min) and delivery followed, although assistance was needed.

Neal (1956) described a two stage caesarean operation for cases of hydrallantois and, in a modified form, this technique is still used at Liverpool. A left flank laparotomy as for a caesarean operation is performed and the uterus identified - usually it lies just inside the incision. A purse string suture is laid through the uterine wall in a circle of about 5 cm diameter. A stab incision is made through this into the uterine lumen, and a large bore sterile tube (e.g. stomach tube) with holes cut over several feet from the tip pushed through to drain off the allantoic fluid. The purse-string suture is tied tightly round the tube to keep it in position. As the fluid is drained away, the pulse rate is carefully monitored. If it shows any tendency to accelerate then the drainage is stopped for 10-15 min before proceeding more slowly. This procedure prevents splanchnic pooling developing to a severe degree. When as much fluid as possible has been drained off, the tube is withdrawn and the purse string suture pulled closed. A routine caesarean operation is now performed immediately, rather than being delayed 24 hours as originally described.

During the post-operative period, these cows consume vast quantities of salt, and mineral licks for the cows to eat (sic!) should be provided for 3 or 4 days. The placenta is often retained (see page 165). The cow should be milked daily for a week, during which time milk usually appears in the mammary gland and the cow will often go on to complete a successful lactation even though the pregnancy may have been terminated two months before term.

(ii) The Mare

Vandeplassche and others (1976) have described 8 cases of 'hydrops' in mares, 6 of which were in Belgian Draft Horses. As in cattle, the condition developed in the last third of gestation and was characterised by progressive abdominal enlargement with diminished appetite and difficult defaecation and gait. Rectal palpation was difficult because of the enlarged uterus. These authors found that rupture of the allantochorion at the cervix produced a marked improvement. In some cases this happened spontaneously and in others could be achieved by manual dilation of the cervix and tearing of the allantochorion (see also p. 165). The allantoic fluid could usually be rapidly siphoned off and the foals then delivered per vaginam. Further evacuation of the uterus, including delivery of the fetal membranes was achieved by oxytocin infusions (see page 165), repeated as necessary.

D. POST-OPERATIVE COURSE

I CARE OF THE NEW BORN

(a) Immediate Attention

(i) It must be ensured that bleeding from the umbilical cord has ceased and it may occasionally be necessary to grasp a bleeding artery or vein with a pair of haemostats.

(ii) It is also necessary to ensure that the young animal is breathing and has no fluid obstruction or respiration. It may be necessary to hold the young up by its back legs briefly to drain the upper respiratory tract. Tickling the interior of the nasal passages with a straw will often encourage the reluctant animal to breathe.

(b) Later Attention

The second requirement is that the animal should be dried off thoroughly. It is a source of amazement to the author how dry a mother can lick her young and the dam is undoubtedly better at drying the neonate than even a body of veterinary students. However, the dam cannot always be relied upon to perform this task after a caesarean operation and attendants should be encouraged to dry the young with wisps of straw with the dual objective of drying it and stimulating it.

The third stage of care is to ensure adequate immunological protection. Where the dam has adequate milk some of this may be taken off by hand and adminstered to the young by stomach tube. In the dam which is not providing milk, maternal blood should be collected, the plasma separated off by centrifugation if possible, and given to the young. It is the author's practice to administer a single injection of an appropriate antibiotic to the young and to spray the navel with an antibiotic plus crystal violet.

In the suckler beef unit and in the case of mares and ewes, the establishment of a bond between mother and young is crucial and the two should be introduced to each other as soon as possible and feeding by the young encouraged. Considerable patience may be required, especially in mares.

II CARE OF THE DAM

(a) Getting the dam to her feet

The cow which was recumbent at the time of the operation or became so during surgery, should have a rope placed between her hind legs just above the hocks so as to prevent her from 'doing the splits' if she tries to rise. The development of milk fever must be carefully watched for and the cow left, preferably in a loose box with plenty of straw, until she is able to stand freely. The calf is best put with the dam immediately.

The ewe may be given any live lambs immediately after surgery is completed.

The mare is best left to recover from general aneasthesia by herself and the foal, if alive, introduced when the mare is standing well enough not to be a danger.

The sow is best returned to her farrowing crate whilst still anaesthetised and the piglets put with her.

(b) The normal course

The prophylactic use of antibiotics and the role of ecbolics in reducing uterine size have already been discussed (pages 153 and 160).

Systematic antibiotics cover should be continued for between 4 and 7 days as appropriate to the antibiotic used and the state of health of the dam.

The dam may appear ill for 16-24 hours post operatively and a small proportion will die, apparently of shock, during this period.

The fetal membranes may be passed spontaneously in the first twenty four hours, especially if the young was alive, but they are usually retained for some time - see under (c) Metritis.

The dam should be fed good feed during the immediate post-operative weeks to enhance recovery.

(c) Metritis

Metritis following dystocia is usual and underlines the need to place antibiotic pessaries in the uterus after removal of the young. The metritis is especially severe if the caesarean was elective with a closed cervix.

(i) The Cow

The fetal membranes are generally delivered within 12 hours of surgery. Daily vaginal examination should be carried out to determine whether retained fetal membranes can be removed easily and what can be removed easily should be removed.

Vandeplassche, Bouters, and others (1974) state that in the cow beyond 4 days, local antibiotic therapy is effective in controlling metritis in only about half the cases. They, therefore, recommend the intramuscular administration of 20 mg of oestrogen on day 5, _careful_ removal of fetal membranes and uterine lavage daily for the following 4 days, by which time uterine involution is usually complete. The author's experience vindicates the value of this procedure. This positive approach to post-parturient metritis should be extended by re-examining the animal's uterus at 6 weeks and at 10-12 weeks post-partum and instituting appropriate therapy.

(ii) The Mare

Immediately after completion of surgery in a case with a closed cervix, a vaginal examination should be carried out and an attempt made to penetrate the cervix and rupture the allanto-chorion. The author's experience has been that with a mare under general anaesthesia it is usually possible to pull the allanto-chorion nearly to the vulva and puncture it with a scalpel.

Oxytocin should be given by intravenous infusion once uterine suturing is complete and the allanto-chorion has ruptured (see page 160).

The following day, vaginal and rectal examination should be repeated to assess uterine size and presence or absence of fetal membranes. A further infusion of oxytocin may be required to assist in evacuation of the uterus.

Edwards, Allen and Newcombe (1974) have suggested that infection develops even if antibiotic pessaries are placed in the uterine lumen after surgery, and that such infection may require several oestrous cycles before it resolves. Systemic post-operative examinations as for the cow are of value.

(d) Wound Breakdown

Repair of the laparotomy wound was discussed on page 159 ff.

In cattle, emphysema of the flank may develop if the peritoneum is not sutured. Little can be done about its development and it usually remains limited to the area immediately around the wound and regresses spontaneously.

Sometimes serum-like fluid accummulates between the muscle layers or under the skin, especially if dead space was left unsutured. This usually resolves best if left alone.

Occasionally abessation occurs and an abscess can usually be lanced and drained easily by removing the lowest skin sutures, penetrating with a gloved hand to the fluid and establishing drainage. If the wound is kept open for several days and irrigated, second intention healing usually follows.

Oedema is the usual sequel to a laparotomy in a mare, especially if the flank approach is used. When the operation is done in the mid-line the oedema gravitates cranially along the body wall and can be readily dispersed by gentle exercise. The fact that the mare's uterus can usually be readily exteriorised, particularly through the mid-line, means that wound contamination is less of a problem than with the cow.

(e) Peritonitis

Some degree of peritonitis is inevitable, especially if the uterus was infected at the time of operation. Although Swanwick and Milne (1973) have shown that infection alone is not capable to inducing peritonitis - some foreign material is essential - the intraperitoneal injection of antibiotics on the second post-operative day in cows from which emphysematous calves have been delivered (see page 161) appears imperative.

In cattle, diarrhoea is the most obvious outward sign of peritoneal inflammation although the cow will also be depressed, pyrexic and inappetent, as well as having a tucked-up appearance. Adhesions begin to develop after about 4 days and they and the omentum can often wall off an inflamed area and so limit the extent of the peritonitis. However, infection can sometimes break out of such a walled-off area and precipitate another crisis which again resolves by being walled off. This cycle of containment and subsequent breakout may go on for weeks leaving the cow weak and emaciated, never to recover. It underlines the need to treat cases which may develop peritonitis intensively during the first few days after laparotomy (see page 153).

Some have advocated that adhesions which form be broken down at about 4 days at rectal examination. However, it seems unlikely that this prevents adhesions forming later, and Sloss and Dufty (1977) observed apparent spontaneous resolution of adhesions during later pregnancies. Other authors have advocated the intra-peritoneal infusion of polyvinyl pyrrolidone (2 or 4% in a litre; Bosted and Brummer, (1969). However, a review of the factors which may cause abdominal adhesions or may reduce the frequency with which they occur (Ellis, 1971) suggests that the absence of ischaemic damage to the tissue underlying the peritoneum was much more important than any therapy.

III POST-OPERATIVE FERTILITY

Vandeplassche, Bouters, Spincemaille and Herman (1968) have reported that of 1875 cows and heifers which underwent a caesarean operation, 60% were served or inseminated again and 74% of these became pregnant with an average of 1.8 inseminations per conception. (Similar observations have been made in Germany, quoted by Ahlers, Lumann and Andressen, 1971). 633 of these cows followed through pregnancy and 91% came to term. The thirty six remaining cows aborted in months 4-6 of pregnancy. Vandeplassche and others also noted an increased incidence of hydrallanotois in cows in pregnancies after caesarean operations and an increased incidence of failure of the cervix to dilate at subsequent parturitions. They presented evidence that these abnormalities were due to large scar deformation of the uterine wall which either prevented normal expansion of the uterus or interfered with fetal nutrition or both.

They concluded that it was unlikely that the situation could be improved upon.

These authors' excellent survey made did not relate subsequent fertility to the indication for the caesarean. A small survey from general practice, the other details of which were published by Morten and Cox (1968), showed that fertility was normal following caesarean operations undertaken for fetal oversize with live calves. Thurley (1979) observed that the fertility of ewes the season after hysterectomy to procure minimal disease lambs was the same as that of a control group. It would seem that the state of health of the uterus at the time of surgery is an important factor in future fertility.

E. REFERENCES

Aehnelt, E., Grunert, E. and Andresen, P. (1971) Entwicklung von Auszug, Embryotomie und Schnittentbindung in der Rindersgeburtschilfe des 19 und 20. Jahrunderts. Deutsche Tierartzliche Wochenschrift 78 : 557-592.

Ahlers, P., Luhmann, F. and Adresen, P. (1971) Komplikationen bei der Schnittentbindung des Rindes unter Berucksichtgung von Haufigkeit, Prophylaxe und weitever Fruchtbarkeit. Praktische Tierartzliche 52 : 573-577.

Arthur, G.H. (1957) Some notes on the foetal fluids of ruminants with special reference to hydrop amnii. British Veterinary Journal 113 : 17-18.

Bierschwal, C.J. and DeBois, C.H.W. The technique of fetotomy in large animals. VM Publishing Inc., Kansas.

Bosted H. and Brummer, H.P. (1969) Versuche zur Verhutung intrabdominelle Adhaesionem nech geburtshilfelichen Laparotomien bein Rind mit Hilfe von Polyrinylpyrrolidon. Berliner und Munchener Tierartzliche Wochenschrift 82 : 429-432.

Cash, R.S.G., Ousey, J.C. and Rossdale, P.D. (1985) Rapid strip test method to assist management of foaling mares. Equine Veterinary Journal 17 : 61-62.

Copland, M.D. (1976) Anaesthesia for caesarian section in the ewe: a comparison of local and general anaesthesia and the relationship between maternal and foetal values. New Zealand Veterinary Journal 24 : 233-238.

Edwards, G.B., Allen, W.E. and Newcombe, J.R. (1974) Elective caesarean section in the mare for the production of gnotobiotic foals. Equine Veterinary Journal 6 : 122-126.

Ellis, H. (1971) Post-operative Adhesions. Surgery, Gynaecology and Obstetrics, 133 : 497-511.

Deutscher, G.H. (1985) Using pelvic measurements to reduce dystocia in heifers. Modern Veterinary Practice 66 : 751-755.

Ford, G.E. (1965) Delayed caesariansection in the cow. Australian Veterinary Journal 41 : 362-363.

Frerking, H., Andresen, P. and Geyer, K. (1967) Zur Schnittentbindung beim Rind mit Naht und Klammerung der Hautwunde. Deutsche Tierartzliche Wochenschrift 74 : 636-638.

Harsch, J.A. and Hanks, D.R. (1971) Public symphysiotomy for relief of dystocia in heifers. Journal of the American Veterinary Medical Association 159 : 1034-1036.

Morten, D.H. and Cox, J.E. (1968) Bovine dystocia - a survey of 200 cases met with in general practice. Veterinary Record 82 : 530-537.

Messervey, A., Yeats, J.J. and Pearson, H. (1956). Caesarean section in cattle. Veterinary Record 68 : 564-568.

Neal, P.A. (1956) Bovine hydramnios and hydrallantois - a report of five cases treated surgically. Veterinary Record 68 : 89-97.

Noordsy, J.L. (1979) Selection of an incision site for caesarean section in the cow. Veterinary Medicine/Small Animal Clinician 74 : 530-537.

Oehme, F.W. (1967) The ventro-lateral caesarean section in the cow. Veterinary Medicine 62 : 889-894.

Pearson, H. (1971) Uterine torsion in cattle - a review of 168 cases. Veterinary Record 89 : 597-603.

Swanwick, R.A. and Milne, F.J. (1973) The non-suturing of parietal peritoneum in abdominal surgery of the horse. Veterinary Record 93 : 328-335.

Skydsgaard, J.M. (1965) The pathogenesis of hydrallantois bovis. Acta Veterinaria Scandinavia 6 : 193-207 and 208-216.

Sloss, V. and Dufty, J.H. (1977) Elective caesarean operation in Hereford cattle. Australian Veterinary Journal 53 : 420-424.

Spincemaille, J., Vandeplassche, M and Bouters, R. (1970) Optreden en behandeling van torsio uteri bij de merrie. Vlaams Diergeneeskundig Tijdschrift 39 : 653-662.

Tavernor, W.D. Trexler, P.C., Vaughan, L.C., Cox, J.E. and Jones, D.G.C. (1971) The production of gnotobiotic piglets and calves by hysterotomy under general anaesthesia. Veterinary Record 88 : 10-14.

Thurley, D.C. (1973) The breeding performance of ewes after hysterectomy. New Zealand Veterinary Journal 21 : 102.

Vandeplassche, M.M. (1973) Caesarean section in horses. Veterinary Annual 14 : 73-78.

Vandeplassche, M., Bouters, R., Spincemaille, J. and Herman, J. (1968) Wird beim Rind die Trachtigkeit durch eine vorausgegangene Schnittenbindung beeintrachtigkeit? Zuchthygiene 3 : 62-69.

Vandeplassche, M., Bouters, R., Spincemaille, J. and Bonte, P. (1974) Technik und Erfolg des Kaiserschnittes bei Kuhen mit emphysematosen Kalb. Deutsche Tierartzliche Wochenschrift 81 : 591-594.

Vandeplassche, M., Bouters, R., Spincemaille, J. and Bonte, P. (1974) Induction of parturition in cases of pathological gestation in cattle. Theriogenology 1 : 115-121.

Vandeplassche, M. Bouters, R., Spincemaille, J. and Bonte, P. (1976) Dropsy of the fetal sacs in mares : induced and spontaneous abortion. Veterinary Record 99 : 67-69.

Vandeplassche, M., Bouters, R., Spincemaille, J. and Bonte, P. (1977) Caesarean section in the mare. Proceedings of the Annual Convention of the American Association of Equine Practitioners 23 : 75-80.

Vandeplassche, M., Oyaert, W., Bouters, R., Vandenherde, C., Spincemaille, J. and Herman, J. (1965) Uber die Eihautwassersucht beim Rind. Wiener Tierartzliche Monatsschrift 52 : 461-473.

Vandeplassche, M., Paredis, F. and Bouters, R. (1961) Flanksnede bij der Rechtsstrande of neerlingende merrie voor hed ontdraaien van torsio uteri. Vlaams Diergeneeskunde Tijdschrift 30 : 1-11.

Vandeplassche, M., Paredis, F. and Bouters, R. (1962) Technik, Resultate und Indication des Kaiserschnittes beim Pferd im Vergleich zur Fetotomie. Wiener tierartzliche Wochenschrift 49 : 48-61.

Vandeplassche, M., Spincemaille, J. and Bouters, R. (1971) Aetiology, pathogenesis and treatment of retained placenta in the mare. Equine Veterinary Journal 3 : 144-147.

Chapter Nine
TRAUMA TO THE FEMALE REPRODUCTIVE TRACT

A. TRAUMA TO THE CAUDAL REPRODUCTIVE TRACT

I INTRODUCTION

The vestibule is that part of the female reproductive tract between the urethral orifice and the vulva. The vulva is the opening of the vestibule at the skin-mucous membrane junction (see p.127).

This section is concerned with trauma to the vestibular and perineal areas and to the caudal, retroperitoneal portion of the vagina. (Trauma to the more cranial portions is considered in the next section).

Lacerations in this area may follow delivery by forced traction of an oversize fetus or, especially in the mare, spontaneous delivery of an abnormally postured fetus. Occasionally, injuries are caused sadistically or, in mares especially, at service by vigorous stallions.

Injuries are conveniently classified into four types:-

First degree lacerations involve only the mucous membrane of the vestibule and/or vagina and/or the vulva. It is convenient to include here contusion without discontinuity and especially haematoma of the underlying tissues.

Second degree lacerations involve the full thickness of the reproductive tract wall but do not involve the rectal wall or the anus. Such lacerations usually occur laterally, especially at the vagino-vestibular junction and may or may not be accompanied by severe haemorrhage from a lateral vaginal artery or vein.

Third degree lacerations involve reproductive tract and rectal walls in toto. If the perineum and anal sphincter are involved, the laceration is complete and should be called a recto-vestibular tear (not a recto-vaginal tear - most such tears and fistulae are in the vestibular portion of the tract). Sometimes a third degree laceration occurs without the perineum and anal sphincter being involved - the defect is then termed a recto-vestibular fistula. Such a fistula may occur if a fetal malposture which resulted in a foot being pushed through vestibular roof (usually to the vagino-vestibular junction) is corrected before the perineum is torn or it may develop if a recto-vestibular tear undergoes some healing. Some fistulae may be so small as to be missed on cursory examination and may only be detected by the presence of faecal material in the vagina/vestibule and around the cervix of the well-conformed mare. Careful and clean examination of all lacerations is imperative to assess the exact location, size and depth of the tear.

II FIRST DEGREE LACERATIONS

Many minor lacerations produced at parturition require no treatment, but the danger is that Fusiformis necrophorous, Corynebacterium pyogenes or anaerobic infection will establish itself, resulting in abscess formation or a severe necrotic vaginitis. Either condition causes severe, unremitting tenesmus and exhaustion and may produce toxaemia. It is preferable, therefore, to treat all contusions with antiseptic cream locally and parenteral antibiotics. Monitoring the progress of the white cell count will enable the development of an abscess to be detected.

Haematoma should not be touched and often regress. Surgical drainage under 10 days is not advisable and then the incision should be made vertically not horizontally, not only to allow adequate drainage but to avoid incising across vestibular constrictor muscles. It will be necessary to keep the drainage incision open long enough to ensure that granulation is proceeding satisfactorily. Dalahunty (1968) has argued that haematoma in the mare should be aspirated (not incised) and irrigated with a chymotrypsin solution once or twice daily for 3 or 4 days. Abscesses may develop in untreated haematoma and will be characterised by marked straining and by the space-occupying mass. They can be drained via the vaginal wall, the pelvic fascia or even (Vaughan 1978) the inguinal region.

III SECOND DEGREE LACERATIONS

Post-partum discharges of necrotic debris and malodorous pus that are not characteristic of retained membranes may be the first sign of trauma if the animal has not been examined post-partum.

Where possible second degree lacerations should be sutured but it is not usually easy to do so and many heal spontaneously provided severe secondary infection is controlled by application of antiseptic creams or careful flushing with weak (0.5%) iodophur solutions. Parenteral antibiotics are indicated. Pain may prevent normal micturition so adequate urine drainage must be provided. A complication of such lacerations in fat heifers is prolapse of fat through the tear. Ligation and excision and loose suturing are indicated.

Severe haemorrhage from vaginal vessels must be dealt with promptly. If an owner should ring stating they have a cow bleeding, they should be advised to put their hand in the cow to locate the bleeding point. If blood is spurting from a point they should be advised to put a finger on it. A pair of haemostats should be clamped on the severed end of the blood vessels for immediate control of haemostasis. It is preferable (G. Wyn-Jones, personal communication) to place two ligatures round the blood vessel proximal to the haemostat rather than trust the haemostat alone - even when the haemostat is removed at 24 hr, there is a risk of reactionary haemorrhage if a ligature has not been applied.

Other possible sources of post-partum haemorrhage must be distinguished from that from vaginal vessels. They include haemorrhage from umbilical vessels and uterine vessels. A rush of blood immediately after deliver of the fetus is likely to be due from the severed ends of the umbilical vessels. In cattle this may occur in cases of uterine inertia. In mares it may occur in cases in which the umbilical cord has ruptured prematurely - normally, the foal remains attached to the placenta for several minutes after delivery. Such haemorrhage does not, of course, affect the mother, and rarely lasts very long.

IV THIRD DEGREE LACERATIONS

Spontaneous healing and resolution of third degree lacerations can and does take place. This tendency is probably the reason, if not the sole reason, for the success often claimed by those who operate upon such lesions immediately after they are formed. However, at the time of injury the tissues are friable and often oedematous and it is not always possible to tell which tissue is alive and which are dead. In such a situation healing occurs in spite of, not because of, surgical interference.

No attempt, therefore, should be made to repair a third degree laceration until sufficient time (up to 6 weeks) has elapsed for granulation and natural resolution to have taken placed. At the time of parturition, only first aid measures are necessary, e.g. haemorrhage control, cleaning up, tetanus anti-toxid or vaccination booster in mares. Placental retention should be dealt with appropriately. In the case of a mare, the owners should be advised that it is unlikely that the mare can be covered under 10 weeks and that it would be better to leave the mare until the following year.

The concept of leaving third-degree lacerations for about 6 weeks is derived from that of Aanes (1964, 1973) who also originally described the two-stage operation for repair of such lacerations, a modification of which is described below.

(a) Preparation for Surgery

The author prefers to keep the mare off hay or grass for 48 hr prior to surgery, during which time, sloppy bran mashes are fed and water is allowed ad libitum.

The mare is restrained adequately in stocks and given a tranquillizer - xylazine given slowly intravenously at a dose rate of 0.5mg/kg is excellent, although the dose will need to be repeated after about half an hour. Epidural anaesthesia is the anaesthetic method of choice (see Green and Cooper, 1986). Local infiltration of the operative area with 3% lignocaine plus adrenaline and administration of 0.5 mg morphine sulphate/kg intravenously just before surgery commences is also effective.

The tail is wrapped and held to one side by an assistant and the rectum is emptied manually as far forwards as possible. The rectum, vagina, vestibule and vulva are all cleaned as for surgery with a mild antiseptic solution. The vestibular wall may be spirited but the rectum should not as the spirit is irritant and makes the rectal glands produce vast quantities of mucus.

Either stay sutures are placed at the levels of the ventral part of the anal sphincter and the upper commisure of the vulva to increase exposure of the surgical field or assistants with large tissue forceps hold the lips of the rectum and vestibule apart.

(b) Stage One Operation

The aim of this first stage operation is to re-create the shelf between rectum and vagina/vestibule - it is of no use to try to restore completely the anatomy of the perineum at one operation as was the aim of earlier techniques. As with most plastic surgery the procedure is to create fresh edges of tissue which can be brought into apposition and allowed to heal together.

If the mare has a recto-vestibular fistula, this will usually need to be converted to a third degree laceration before proceeding simply by incising the existing shelf in the mid-line with a pair of scissors or a scalpel. In cases where the fistula is small, it may be possible to close it by incising between rectal floor and vaginal roof along the fistula's periphery and suturing the exposed surfaces across the hole, but this is rarely worth the difficulty as there is little risk that the existing shelf will not heal if incised.

Examination of the perineum will reveal a broad scar at the junction of rectal (brick red) and vagino-vestibular (pink to white) mucosae and this must be incised along the length of its junction with vagina vestibule, i.e. extending along the left-hand wall of the common opening of rectum and reproductive tract, across the shelf formed cranially where rectum and reproductive tract are intact and along the right hand wall of the common opening.

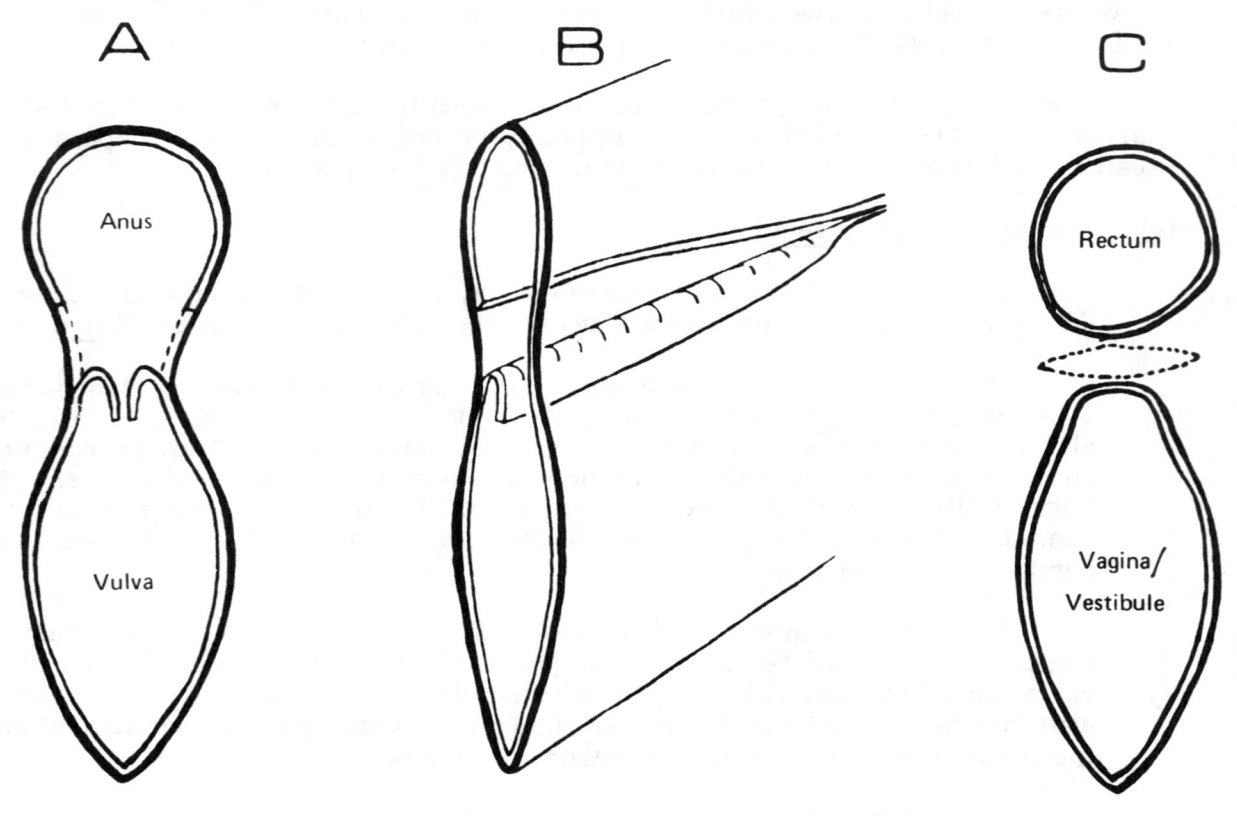

Fig. 9:1 Showing the dissection stage of repair of a third-degree perineal laceration. 'A' represents the external view and 'B' the right latero-caudal view. 'C' represents a section at the level where the recto-vaginal shelf is still intact, dissection opening up the area between as shown by the dotted lines. (After Aanes, 1962)

A flap of mucous membrane is then dissected away from this line ventrally for about 3 cm to expose perivaginal fascia (Fig. 9:1). Beyond the cranial limit of the lesion the rectal floor and vagino-vestibular roof are separated for a distance of about 5 cm. This dissection may be difficult if a marked fibrous tissue reaction has occurred, but haemorrhage is usually slight. If too thick a flap is taken then the vaginal or vestibular vessels may become involved and haemorrhage will be profuse.

Using interrupted suture material on a cutting needle (the author prefers 3 B.P. Chromic catgut swaged onto a cutting needle), 2 or 3 interrupted sutures are placed in the perivestibular fascia at the cranial end of the shelf between rectal floor and reproductive tract roof (Fig. 9:2). A variety of suture patterns may be employed - the simple one shown here, a figure of 8 or a 6 bite pattern taking in rectal submucosa, vestibular wall, peri-vestibular fascia and vestibular submucosa on each side. These are the most difficult sutures to place but they are the most important. Each suture is tied immediately after it is placed forming a shelf which will not allow passage of faecal material from rectum to vagina. This line of interrupted sutures is continued caudally to the perineum (Fig. 9:2). Careful attention is paid to providing a thick shelf and to placing sutures accurately. The size of the anal opening must not be unduly restricted at this stage or severe straining will result and the sutures may tear out - the anal sphincter is, therefore, NOT sutured.

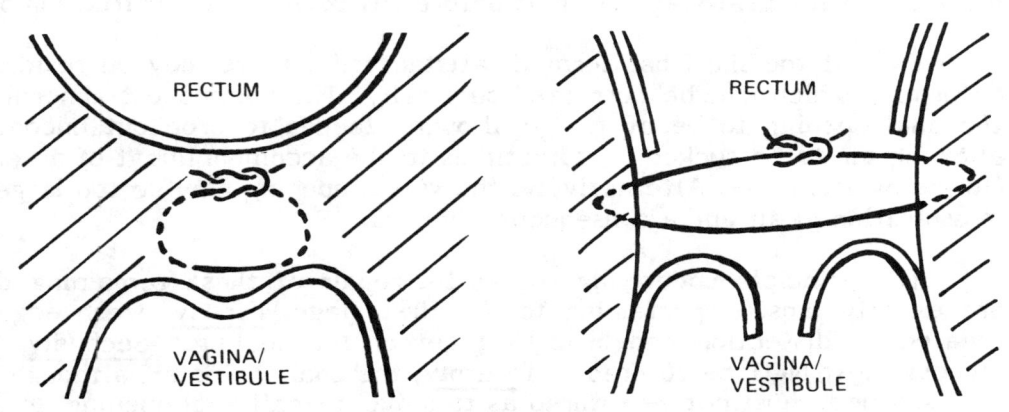

Fig. 9:2 Showing the suturing stage of a repair of a third degree perineal laceration. The left hand figure illustrates the repair where the recto-vaginal shelf is still intact (compare 9:1c). The right hand figure illustrates the repair where the shelf is not present and where a new shelf is created. (After Aanes, 1964)

It is important that none of these sutures involve the rectal mucosa which is is very sensitive to implanted suture material. The resultant irritation makes the mare strain and reduces the likelihood of the repair healing. It is also important that none of these sutures penetrate through vestibular mucosa, so forming a track for infection through the membrane.

An additional suture line which may be laid is a continuous suture along the ventral edge of the reflected flap of vestibular mucous membrane. This serves as an additional seal between rectum and vestibule the author has never done it and cannot attribute his failures to his not using it.

Post-operatively the mare is fed bran mashes only for two days and kept on a fairly laxative diet for about a week (grass is best) so that her faeces do not become too hard. Straining by the mare during the process of healing of the wound must be avoided - hay, therefore, is contra-indicated for the first few days.

Tetanus anti-toxid or vaccination boosters should be administered and a course of 5 days of antibiotic therapy instituted, the first injection having been given before surgery commenced.

The tail bandage will need changing and the perineal area cleaning daily.

After not less than a week, the wound may be inspected by careful vaginal examination. At this time a good indication can be obtained as to whether or not the shelf has been re-created. If non-absorbable sutures have been used they should be removed now. It is, not, however, possible to judge how good the final cosmetic appearance will be for several weeks - the animal's capacity to remodel the area following surgery is amazing.

(c) Further Surgery

The mare should be inspected again about 4-6 weeks after the first stage operation to determine (i) whether a satisfactory shelf has been formed and (ii) whether it is necessary to undertake any perineal reconstruction.

If the shelf is not complete, e.g. a fistula remains, then attempts should be made to create a satisfactory shelf before any perineal reconstruction begins.

Even if the shelf has formed satisfactorily, there may be residual perineal deformity which can be corrected surgically. The commonest abnormality is for the anal opening to be overlarge allowing faeces to drop out uncontrolled and allowing air to be sucked in, sometimes to the accompaniment of an embarassing (to the owner) noise. Alternatively, the vulval opening may be too large, resulting in aspiration of air and a consequent vaginitis.

The principle underlying successful repair of these remaining deformities are exactly those appertaining to the first stage. Firstly, fresh edges must be created by dissection and brought together for healing. Secondly, the rectal mucosa must NOT be sutured and thirdly, the anal sphincter, although edges may be freshened, must not be sutured as to make a small anal opening. In addition, it is important that the rectal lumen should not be suddenly narrowed at the anus but reduced gradually.

Aanes (1973) has described a technique for repairing an anal sphincter but this should only be attempted when all other repair and remodelling is satisfactory. It is re-emphasised, however, that in the majority of cases extensive remodelling takes place.

At least one oestrus should be allowed to come and go after the repair has been completed to allow any endometritis to resolve before the mare is covered.

(d) Prognosis

The minimum realistic interval between foaling and first possible service after repair is ten weeks. Unless the mare has foaled particularly early in the season and the client is prepared to have a late foal the following year, then it is preferable to have a whole season in which to attempt surgical correction and allow remodelling of the area. The owner should always, therefore, be advised at the time that the damage occurs that it is preferable that no attempt be made to get the mare in foal again that year.

At subsequent foalings, there is a slightly increased risk of second degree lacerations because of a lack of elasticity but a further third degree laceration is unlikely. Any damage that does occur can be dealt with by appropriate measures.

(e) <u>Other Species</u>

Although the above remarks have been made specifically about the mare, they apply equally to the cow. Indeed, in the cow spontaneous resolution and remodelling occur remarkably well. In the sow and ewe it is unlikely that the value of the animal will be such as to justify the expense of remedial surgery.

B. TRAUMA TO THE CRANIAL REPRODUCTIVE TRACT

Tears of the uterus, cervix and cranial vagina are uncommonly recorded complications of late pregnancy, occurring primarily in cattle and sheep. The glans penis of the stallion can perforate the cranial vagina of the mare and this condition is discussed below (p.180).

I PREDISPOSING FACTORS

The location of tears is usually the dorsal surface of the gravid horn and/or body and may occasionally extend caudally into the vaginal roof. They may be transverse (especially in uterine torsion - see Ferny and others, 1961) or longitudinal. Exceptionally, a hole may develop at the tip of the uterine horn (Vandeplassche, personal communication).

The available evidence suggests that the great majority of cases are associated with and probably result from dystocia.

Some result from <u>assisted delivery</u>. Uncaring lay or veterinary obstetrical manipulations may result in tearing of uterus at attempted delivery and this is especially true of the fetus has a malposture or malpresentation and more likely in sheep than cows. In all cases with a tear then the attendant should be asked if he has had his hand in. If he has produced a tear he should be told so, now!

Other cases are <u>spontaneous</u> i.e. not iatrogenic, although associated with some fetal or maternal factor. The three most common conditions are (i) uterine torsion, (ii) fetal oversize with posterior presentation and especially if 'breech' (bilateral hip flexion), (iii) a male fetus - 89% of bovine cases (Pearson, 1975) and (iv) fetal deformity such as torticollis, ankylosis, muscle contracture, Schistosomus reflexa or severe emphysema. Fischer and Phillips (1986) record two cases in mares which had been treated by intra-uterine infusion for placental retention, but they believed the tears occurred prior to infusion.

Hopkins and Amor (1964) suggest that the fetus becomes impacted in the pelvis and that continued uterine contractions increase intra-uterine pressure to the point where rupture occurs at a weak point - 60% of bovine cases occur in first pregnancies (Pearson 1975) in which the uterus may be overstretched.

The fact that in the majority of cases a predisposing cause can be found suggests that abnormally violent movement on the part of the fetus is not important. Some cases are associated by the client with an accident, such as a fall, but when one bears in mind how many cows may fall in late pregnancy and how rarely the uterus ruptures, it is not easy to prove that the fall was the cause of the tear.

II SEQUELAE AND CLINICAL PICTURE

A wide variety of clinical pictures may result from uterine rupture, some of them radically different from each other. Five general categories may be distinguished.

(a) <u>Incidental Finding</u>

If the uterine tear occurs whilst the cervix is closed and the cervix remains closed, and if no fatal haemorrhage develops then the health of the dam may continue unimpaired and the tear remain undetected. The length of time for which the fetus will remain alive is probably short (not normally exceeding 12 hours) because contractions of the now empty uterus reduce the placental surface area and may actually occlude the umbilical vessels. Pearson (1975) suggests that, even if infection is absent, an extensive fibrinous peritonitis (see (d) below) develops as a result of fetal and placental autolysis. However, Dennis (1966) has recorded a case of abdominal retention in a ewe of at least twelve months duration which was discovered at shearing.

(b) <u>Maternal Haemorrhage</u>

Some haemorrhage at the time of tearing is inevitable but it is rarely fatal (Pearson, 1975). If the fetus is released into the abdominal cavity or is delivered, rapid involution of the uterus controls haemorrhage and the effects are only transient and may not even be apparent.

<u>Transient haemorrhage</u> will be characterised by illness of about 24 hr duration with shivering, anorexia, reluctance to stand or move, pallor and a moderate degree of tachycardia, all of which progressively improve. <u>Serious haemorrhage</u> will also be characterised by pallor and tachycardia but these will become progressively worse.

Maternal cotyledons alone can be torn and bleed and the uterus fill up with blood. The clots should be scooped out (it is usually impossible to find the bleeding point) and the uterus made to involute with ecbolics.

(c) <u>Evisceration</u>

Intravaginal prolapse of intestines is an uncommon sequel to uterine or vaginal rupture, although its occurrence is almost epidemic proportions in extremely fat and well-fed heavily pregnant ewes has been reported by White (1961). The ewes seen by White were almost invariably found dead and Pearson (1975) notes that in both the cases he saw in cattle, the tear in the vagina could not be repaired and the cows had to be destroyed.

(d) <u>Peritonitis</u>

The development of peritonitis, particularly if recognition is delayed, makes for a poor prognosis.

Most cases at term, particularly if delivery is assisted, are grossly infected with <u>Corynebacterium pyogenes</u> and/or coliform or clostridial organisms and a putrid or emphysematous fetus is present - severe peritonitis is inevitable.

Peritonitis in cattle is often associated with rather a vague clinical picture and the animal is not always as severely ill as one would expect from knowledge of peritonitis in other species (see Blood, Henderson and Radostits, 1979) - peritonitis should always be considered in the differential diagnosis of recumbency around parturition. There is usually anorexia, dullness and some degree of ruminal and intestinal stasis develops (although diarrhoea may be present early in the disease). The temperature may be sub-normal and the animal reluctant to rise. Very occasionally vomiting may develop and in some cases abdominal distension is marked due to inflammatory exudate or pneumo-peritoneum from gas-forming organisms.

The possibility of a vagino/uterine tear should be considered. In any mare with post-foaling colic, especially if abdominal sounds are normal, the rectal temperature is rising and tympany is developing.

(e) Cervical lesions

Cervical lesions occur mostly in mares and are not usually noticed until the mare is covered (see page 181). They can produce an incompetent cervix and therefore an endometritis which may result in failure to conceive or abortion.

III TREATMENT

If the tear is small and rapid involution of the uterus by oxytocin injection or infusion (see page 160) can be achieved, then no treatment is required other than systemic antibiotic for a suitable period. It seems likely that many tears in sheep heal without any treatment - see Hindson (1966). Mares should, however, be cross-tied.

(a) Repair of the tear per vaginam

Several authors report extensive series of cases repaired in this way in cattle. Caudal epidural anaesthesia is the anaesthetic of choice for the cow and general anaesthesia for the rare cases in the horse. Absorbable suture material is used. The suturing usually has to be done single-handed, especially if the tear extends far cranially, but the knots may be tied outside the vulva and slid along the suture material. A continuous suture is satisfactorily and the aim should be to invert the edges of the tear into the uterine and/or vaginal lumen. Some authors report everting the uterus to effect repair rather than resort to laparotomy.

The disadvantages of the vaginal approach is that it does not allow an adequate inspection of the peritoneal cavity and an assessment of the extent, of or likelihood of peritonitis, but it is easier to repair a vaginal tear or a tear in the uterus close to the cervix or one in a small contracted uterus by this route.

(b) Repair of the tear at laparotomy

Laparotomy is positively indicated when the fetus is partly or completely in the peritoneal cavity so that delivery can be effected.

In cattle, paravertebral anaesthesia (Appenidix IV) of the left flank may be employed whilst in the mare general anaesthesia and a mid-line approach would be the author's preference.

Laparotomy also allows an attempt to be made to break down adhesions and to remove bits of placenta and fetus. Peritoneal lavage may be of benefit.

If the uterus has involuted, then it can be difficult to suture it as it retracts into the peritoneal cavity - the sloping incision described for caesarean operation in cattle (page 154) is advantageous here.

Pearson (1975) records that particular difficulty was experienced in repairing extensive tears of the vaginal roof because the tears were not always in the same plane of tissue at different depths.

(c) Haemorrhage

When recognised, the simpler cases of haemorrhage are best treated by repair and rapid involution.

Blood transfusions without ligation of the point of loss or marked contraction of the uterus can sometimes be counter-productive as they raise the blood-pressure and this may reduce clotting ability. The value of blood transfusions must be matched against their potential disadvantage and, in general practice, the practicality of giving reasonable amounts - the average 500 kg cow can generally survive the loss of 20 litres of blood without replacement.

(d) Evisceration

Whether such a case can be treated or not, depends not only upon the length of time elapsing between prolapse and its recognition but also upon the facilities and level of assistance available. As with prolapse of intestines after castrations (page 83), the intestines should be cleaned, preferably with warm saline, any dead or dying lengths removed and an anastomosis performed, before returning the viscera to the abdomen.

(e) Peritonitis

Peritonitis recognised at laparotomy may be treated on discovery by lavage, followed by large doses of soluble antibiotic such as crystalline penicillin intraperitoneally at surgery and two days later.

When peritonitis is suspected later, then again high doses of intraperitoneal antibiotic are indicated. In cattle, peritonitis tends to become rapidly localised and unless treated before that stage the prognosis is virtually hopeless (see page 166). Laparotomy and lavage may be indicated in valuable animals.

C. VAGINAL PERFORATIONS

A rare injury in the mare is vaginal perforation at mating though a similar lesion can be produced sadistically. Although the condition is well known, three series of cases have recently been presented in the literature (Held and Blackford, 1984; Blue, 1985; Tulleners, Richardson and Reid, 1985).

Most perforations occur in the dorsal wall of the cranial vagina and are presumed to be due to a smaller stallion having to thrust dorsally in a larger mare, though Blue (1985) suggests that mating technique may play a part. The lesion should be suspected of blood is discharged from the mare when a stallion dismounts and it can be recognised by careful, clean palpation or with a speculum.

The major risk is from peritonitis brought about by the combination of the semen ejaculated into the peritoneal cavity and the contaminating bacteria (bacteria alone will not usually cause a peritonitis). Prompt antibiotic therapy is, therefore indicated.

In some cases, bowel may eviscerate either just into the vagina or out through the vulva. Prompt resection of devitalised bowel is called for, as is the return of healthy bowel to the abdomen.

The vaginal tear will usually heal by second intention in 7-10 days but the mare should be cross-tied to keep her standing for 48 hours at least. The tear may be sutured per vaginam if desired.

Provided treatment for peritonitis is started early enough and is maintained at an adequate level, the prognosis is good. In some countries it is standard practice to use a breeding roll (which prevents the stallion achieving complete intromission) to prevent the occurrence of vaginal perforation.

D. CERVICAL SURGERY

Difficult births and abused, assisted deliveries may result in prolonged atony or tearing and laceration of the cervix. Healing may occur and normal function may be recovered. The cervix may, however, become a permanently incompetent valve, incapable of maintaining pregnancy - placentitis and early abortion are the usual sequelae.

Surgical techniques for dealing with cervical injuries have been described by Evans and others (1979) and by Brown and others (1984). For anaesthesia see Gadd and Kumar (1980).

The procedures are essentially for those of any plastic surgery in that fresh edges are created and sutured together. The major problem is anaesthesia and access. Suturing of mucosa and sub-mucosa should be with different rows. Defects that extend the full length of the cervix or cause deformities greater than 45° of the circumferential measurement carry little hope of successful repair.

E. REFERENCES

Aanes, W.A. (1964) Surgical repair of third-degree perineal laceration and recto-vaginal fistula in the mare. Journal of the American Veterinary Medical Association 144 : 485-491.

Aanes, W.A. (1973) Progress in recto- vaginal surgery. Proceedings of the 19th Annual Convention of the American Association of Equine Practitioners 225-240.

Arthur, G.H. (1975) Veterinary Reproduction and Obstetrics. Fourth Edition, pp. 288-293. Bailliere and Tindall, London.

Blood, D.C., Henderson, J.A. and Radostits, O.M. (1979) Veterinary Medicine. Bailliere Tindall, London.

Blue, M.G. (1985) Genital injuries from mating in the mare. Equine Veterinary Journal 17 : 297-299.

Brown, J.S., Varner, D.D., Hinrichs, K. and Kenney, R.M. (1984) Surgical repair of the lacerated cervix in the mare. Theriogenology 22 : 351-359.

Dalahunty, D.D. (1968) Surgical correction of contributory causes of uterine disease in the mare. Journal of the American Veterinary Medical Association 153 : 1563.

Dennis, S.M. (1966) Abdominal retention of a dead foetus in a ewe following spontaneous rupture of the uterus. Veterinary Record 78 : 326-327.

Evans, L.H., Tate, L.P., Cooper, W.L. and Robertson, J.T. (1979) Surgical repair of cervical lacerations and the incompetent cervix. Proceedings of the Annual Convention of American Association of Equine Practitioners 25 : 483-486.

Ferny, J., Etienney, J. and Francois, J. (1961) La rupture spontanee de l'uterus chez la vache a terme. Bulletin de Societe de Science Veterinaire de Lyon 63 : 185-191.

Green, E.M. and Cooper, R.L. (1986) Continuous caudal epidural anaesthesia in the horse : an update. Proceedings of the Annual Convention of the American Association of Equine Practitioners 31 : 409-414.

Held, J.P. and Blackford, J.T. (1984) Vaginal perforation after coitus in three mares. Journal of the American Veterinary Medical Association 185 : 533-534.

Hopkins, A.R. and Amor, O.F. (1964) Rupture of the bovine uterus at parturition. Veterinary Record 76 : 904-906.

Pearson, H. and Denny, H.R. (1975) Spontaneous uterine rupture in cattle : a review of 26 cases. Veterinary Record 97 : 240-244.

Tulleners, E.P., Richardson, D.W. and Reid, B.V. (1985) Vaginal evisceration of the small intestine in three mares. Journal of the American Veterinary Medical Association 186 : 385-387.

Vaughan, J.T. (1978) Urogenital surgery. Proceedings of the Annual Convention of the American Association of Equine Practitioners 24 : 543-566.

White, J.B. (1961) An unexplained condition in pregnant ewes. Veterinary Record 73 : 281 and 330.

Chapter Ten
MISCELLANEOUS CONDITIONS OF THE FEMALE REPRODUCTIVE TRACT

A. OVARIECTOMY

I MARE

(a) <u>Indications</u>

(i) <u>Granulosa Cell and other Tumours</u>

The granulosa cell tumour is the commonest type of equine ovarian tumour.

The mare may be presented initially with a history of infertility or with behavioural problems. In either case there may be either absence of oestrous behaviour, or continuous or intermittent and irregular display of oestrus, although in the latter instances such animals may not actually show in-oestrous to a stallion. In the more advanced cases particularly, the mare may show masculine features such as squealing, and flehmen. On rectal examination, an enlarged, hard ovary is palpable the other ovary usually being small and hard. Only occasionally is it possible to palpate the multiple cysts of which the tumour usually consists as the ovarian capsule remains firm and intact. Many such mares have consistently elevated plasma testosterone concentrations (Stabenfelt and others, 1979;)

(ii) <u>Behavioural Problems</u>

Hobday popularised the operation of ovariectomy for mares which had a condition he described as 'cystic ovaries', and it became almost axiomatic that a mare with any behavioural problem had cystic ovaries and required spaying. It is doubtful, however, whether many of the mares he spayed had abnormal ovaries as, at that time, the large size of the normal equine follicle was not recognised (Arthur, 1975). Tubo-ovarian cysts or large paraovarian cysts or large germinal inclusion cysts may be palpable in association with the ovary, but do not seem to affect fertility (Blue, 1985).

A condition characterised by the growth of one or many follicles which do not come to full maturity but which collectively give the ovary a large knobbly feel, is recognised (Arthur, 1975) and sometimes falsely called cystic ovary - the follicle(s) are not persistent. The condition is associated with prolonged display of physical and behavioural signs of oestrus and occurs particularly at the beginning of the breeding season. Treatment is difficult. The injection of human Chorionic Gonadotrophin is often without effect and treatment with a progestagen is preferred.

Occasional mares are particularly troublesome when they are in oestrus and may be such a nuisance at that time as to be useless. Such mares will, of course, cease to be a nuisance when their ovaries are removed.

A third category of behavioural problems is illustrated by a group of mares which continuously show some signs suggestive of oestrus and are, therefore, regarded as nymphomaniacs but, in fact, they fail to accept service. They may be aggressive and bite, and may be unreliable, particularly when handled round the hind quarters. Progesterone and testosterone concentrations are consistently low, indicating absence of ovarian activity. Some but not all such mares also appear to respond favourable to having their ovaries removed, but why they should do so is obscure.

(b) Operative Technique

(i) Preparation

The mare is fully starved for not less than 24 hr prior to surgery to reduce the bowel contents - some surgeons prefer to starve for as long as 48 hr. In order to soften the bowel contents (and so avoid mistaking a hard lump of faeces for an ovary) a laxative diet is fed for 3 or 4 days prior to operation - fresh grass is far preferable to hay. The mare should not be operated upon if she is in oestrus.

(ii) Anaesthesia

The mare is placed in dorsal recumbency under general anaesthesia. An area in the mid-line extending from the mammary gland to cranial to the umbilicus is prepared for surgery. It is especially important to clean the sebum which often accumulates between the halves of the udder.

(iii) Surgery

A mid-line incision is made extending from as far caudal between the halves of the udder as possible, to as far forwards as necessary. If normal ovaries are to be removed, a 15 cm incision is adequate, but if a tumour is present the incision will have to be longer.

A hand is inserted through the incision and the ovary identified. Usually the bladder can be found in the mid-line, then the uterine horns identified just dorsal to the bladder and the ovary found at the end of the horn.

The ovary is brought to the wound. One of the advantages of the mid-line incision is that the ovaries can usually be brought into view and may in some cases even be brought outside the abdominal wound for positive identification. It is not unknown for an intra-abdominal lump to be identified as ovary by palpation and to be removed only for it to be discovered that the lump is a hard ball of faeces in a segment of colon.

Normal sized or small ovaries may be removed using solely an ecraseur. The instrument is threaded with its chain and is opened fully. The operator places one hand through the chain loop, grasps the ovary and then lifts the chain over the clenched had and ovary to encircle the ovarian pedicle. The ecraseur chain is then slowly tightened by an assistant whilst the surgeon ensures that bowel or omentum do not get trapped between chain and pedicle. It is important that the pedicle should not be stretched whilst the ecraseur is tightened or haemostasis will be inadequate, so it is often necessary to proceed with the loop within the abdomen.

In the case of large ovaries, such as granulosa cell tumours, it is often helpful to reduce the magnitude of the pedicle by placing ligatures round those blood vessels nearest the uterine horn, reserving the ecraseur for the major blood vessels. Again it must be emphasised that the encraseur, should be tightened extremely slowly and that the pedicle must not be under tension when this is done. A good ecraseur, with a chain of adequate thickness and total mobility at every joint is essential. The mid-line laparotomy may now be repaired (see page 160).

(iv) Post-Operative Care

It is traditional to keep a very close watch on mares from which ovaries have been removed but experience suggests that they tolerate the removal of normal-sized ovaries as well as male horses tolerate removal of an abdominal testis. Mares which have had tumours removed recover more slowly, but this is probably because more trauma has been inflicted in such cases. Nevertheless, a watch must be kept for signs of pain, of which an accelerating pulse is an indicator, and signs of haemorrhage. As with any operation, infection or wound breakdown may transpire and intra-peritoneal adhesions may develop or become critical only months or years after the ovariectomy.

(v) Prognosis

Mares with granulosa-cell tumours will rarely return to normal cycling activity quickly (Stabenfelt and others, 1979) and up to a year must be allowed. The finding of a hard small contralateral ovary prior to operation must not be taken as a bad prognostic sign.

II OTHER SPECIES

(a) Cattle

Ovariectomy of cattle is now rarely practised in the U.K. but it used to be commonly performed at about 4-6 weeks after partuition for the prolongation of lactation in a cow destined not to be put in calf again. It is still performed on range heifers on the American continent and it can occasionally be justified in an individual cow in which one ovary, because of the occurrence of a tumour or of a persistently luteinised cyst which prevents normal fertility.

As with the mare, the operation should not be performed whilst the cow is in oestrus.

The operation may be performed as described for the mare, under general anaesthesia or deep sedation with xylazine and local infiltration of anaesthetic. Alternatively it may be performed in the standing position under paravertebral anaesthesia (Appendix IV) through the left flank. If the latter technique is employed anaesthesia of the ovarian pedicle must be provided by holding a swab soaked in local anaesthetic solution on the stalk for about 30 sec prior to applying the ecraseur.

The operation can be speedily performed in non-cycling range heifers by using a special instrument designed by Rupp and Kimberling (1982) which penetrates the dorsal wall of the vagina and removes the ovaries in a manner akin to a uterine biopsy device.

(b) <u>Sheep and Goats</u>

The smaller ruminants are rarely presented for a therapeutic ovariectomy, but the operation is simply performed under general anaesthesia through a mid-line laparotomy. Haemostasis can be achieved by a three forceps technique and ligature similar to that employed in a bitch spey.

(c) <u>Pigs</u>

Ovariectomy of the pig is largely of historical interest. Lay people performed the operation on gilts destined to be fattened using a technique similar to that now used to spey cats through the flank.

B. PNEUMOVAGINA

I INDICATIONS

In some mares the vulval lips cease to act as a seal which allows mucus and urine out but, except at breeding, nothing in. In such a case air may be aspirated into the vagina which becomes ballooned and a vaginitis is set up, or particulate matter from the faeces may fall through the unsealed lips into the vagina, again setting up a vaginitis. Sometimes the defect is so considerable that the mare can be heard to suck air into the vagina with a 'fluting' sound.

The vaginitis usually develops into an ascending infection with endometritis and consequent infertility; occasionally urinary tract infection also develops. The level of infertility shown will vary from simply repeated return to service to obvious evidence of endometritis.

Pascoe (1977) has described three types of vulval conformation in mares (Fig. 10:1). The <u>first</u> group have a short (2-3 cm) length of vulva dorsal to the pelvic brim, and although the vulva tends to be drawn forwards with advancing age, the extent is not great. Such mares do not require plastic surgery.

The <u>second</u> group are mares with a longer (6-7 cm) length of vulva above the pelvic brim. With advancing age the dorsal commissure of the vulva is drawn cranially and the angle to the horizontal at which the vulva is inclined becomes small. Such mares require surgery only with advancing years.

The <u>third</u> group are mares which, even as youngsters, have a dished vulval conformation often with the ventral commissure at the level of the pelvic brim. Such mares require surgery for breeding to be attempted, but even after surgery breeding is not always successful, particularly in old age, as the vulva has become too deformed.

The factors that produce these three types are varied. Some young mares in work are kept in too thin a condition and in them the vulva may be drawn well forwards. When they are retired and put on fat, the vulva then reverts to a normal conformation. Pneumovagina can also occur in overweight mares in which the labial lips become inverted. Other causes of poor bodily condition (bad teeth, chronic parasitism) may similarly result in a thin-lipped, toneless muscled vulva which may be drawn too far forward or gape too easily. Trauma at parturition or poor healing of lesions acquired at parturition may likewise affect vulval integrity. In other mares, bad conformation such as a flat croup, or a high tail head may predispose to pneumovagina. Thus, as Walker and Vaughan (1980) point out, pneumovagina is a result of a combination of factors that may occur in young maiden mares or old dams, in the too fat or the too thin, or in those with poor conformation. A solution to the problem may require more than a simple surgical operation.

A further complication of poor confirmation is a change in the angle of the floor of the pelvic canal which can lead to urine draining cranially. Factors involved include multiparity and age which result in the bladder and uterus being pulled cranially over the pelvic brim. The urethral orifice may follow them and this results in various amounts of urine running cranially. Occasionally, the reduced vulval orifice in a Caslicked mare results in urine being splashed back cranially.

It is to be noted that the vulval defect is often more apparent when the mare is in oestrus. Normal stud practice dictates that all barren and maiden mares should be inspected in the autumn and corrective management undertaken and surgery performed, at that point in time.

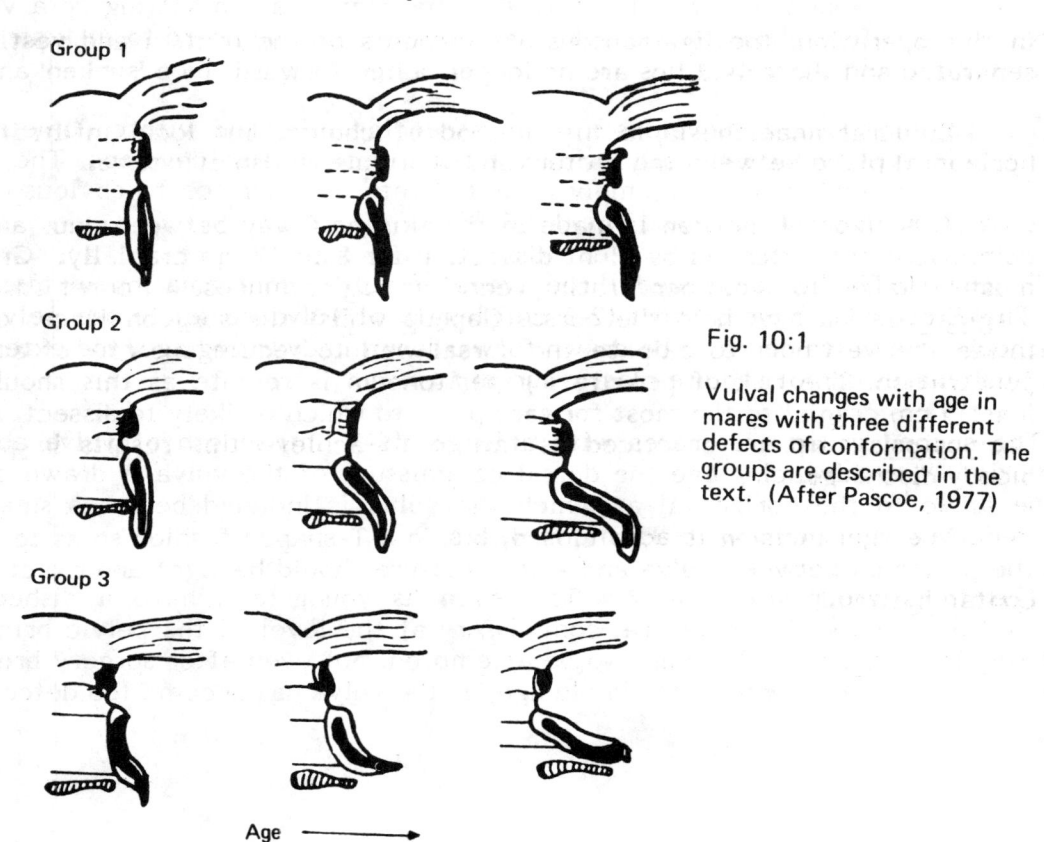

Fig. 10:1

Vulval changes with age in mares with three different defects of conformation. The groups are described in the text. (After Pascoe, 1977)

II OPERATIVE TECHNIQUES

(a) Caslick (1937)

The mare must be adequately restrained and sedated. The use of properly designed stocks and xylazine given by slow intravenous injection (1mg/kg) is ideal but the use of a twitch, raising a foreleg, and operating round a door post are often resorted to. The bandaged tail is held to one side by an assistant and the perineum thoroughly washed and prepared for surgery.

Epidural anaesthesia may be employed but field infiltration with local anaesthetic works well - the use of a dental syringe with its long fine needle has much to commend it. The aim is to anaesthetise a strip of vulval mucosa just inside the mucosa/skin junction about 1 cm wide extending on each side from the dorsal commissure to about 2 cm below the level of the pelvic brim.

The lowermost point of this mucosa is then picked up with rat tooth forceps and a thin vertical strip of mucosa (not exceeding 1 cm in width) is removed from both sides using a pair of scissors, thus creating an inverted-U area denuded of epithelium. The aim is to close the vulval orifice to just below the level of the pelvic brim. Skin should <u>never</u> be removed. Horizontal mattress sutures are then placed so that the denuded surfaces are accurately apposed. The sutures, which may be absorbable or non-absorbable should be placed fairly close together to achieve an effective seal. A protective layer of petroleum jelly is applied over the vulva and tetanus toxoid or a booster vaccination given.

(b) Pouret (1982)

In this operation, the ligamentous attachments of the rectum and vestibule are separated and the vulval lips are no longer pulled forwards by a 'sunken' anus.

Epidural anaesthesia is the method of choice, but local infiltration in a horizontal place between the rectum and vestibule is also effective.

A horizontal incision is made in the skin half way between anus and vulval commisure and extended by blunt dissection for 8 to 12 cm cranially. Great care must be taken to avoid penetrating rectal or vagina mucosa - Pouret advocates a surgical assistant to hold the dorsal flap up whilst the surgeon places one hand inside the vestibule to palpate the dorsal wall to reduce the risk of accidental penetration. The risk of penetrating peritoneum is remote as this should lie at least 5 cm cranial to the most forward point to which is likely to dissect. Absolute haemostosis must be practiced as failure to achieve this results in prolonged bloody discharge.

The skin incision is then closed, but in a T-shaped fashion so as to elongate the perineum between vulva and anus - closure should be tight and exact to avoid contamination.

(c) **Urethral Extension**

Two slightly different approaches for urine pooling have been advocated by Monin (1972) and Brown, Colahan and Hawkins (1978). The latter's technique is now the preferred one, and aims to create a mucosal tube from the urethral orifice as far cranially as possible to enable urine to be directed to the ventral vulval commisure.

The mare is tranquillised and placed in stocks. Epidural anaesthesia is the method of choice. A No. 30 Foley catheter is placed to stop urine contamination of the site during surgery.

The lips of the vulva are retracted and the transverse fold over the urethral orifice is split into dorsal and ventral layers and the incision extended caudally along the lateral wall of the vestibule to within 2 cm of the vulval lips (Fig. 10:2).

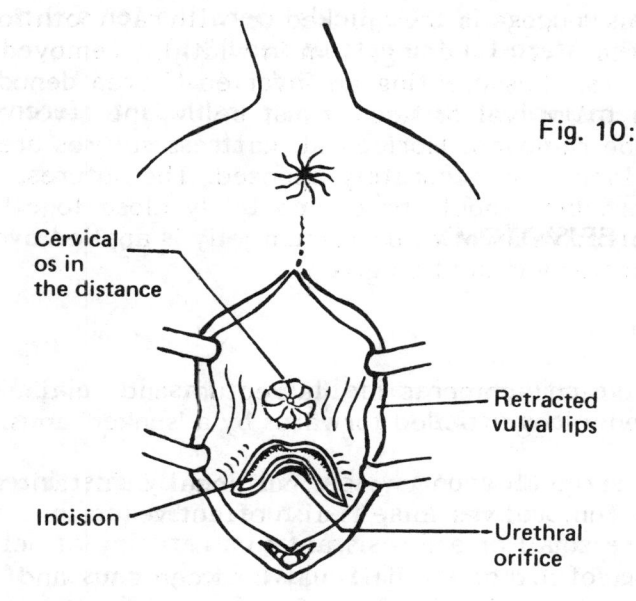

Fig. 10:2 Showing the dissection stage of Brown's operation for urethral extension. The incision has been made through the transverse fold above the urethral orifice, splitting it into dorsal and ventral layers, and then cardally along the lateral wall of the vestibule

The mucosa along the incision is undermined ventrally and dorsally to produce two mucosal shelves (Fig. 10:2) which can be readily drawn across the vestibular floor to form a roof to the Foley catheter.

Repair is in three layers, preferably with absorbable sutures. The ventral mucosa is closed with a continuous suture, aiming to evert edge into the tunnel. The sub-mucosal tissue is sutured with interrupted sutures and the dorsal mucosa closed with another continuous suture. The catheter is removed to avoid the risk of cystitis.

Problems arise particularly in old mares in which the tissue immediately dorsal to the original urethral opening is often too fragile for good closure and fistulae may develop here with continued backflow of urine into the vagina. Vigorous stallions may rupture the repair at service.

III POST-OPERATIVE CARE

Non-absorbable sutures are removed after a week. A period of sexual rest is allowed, the period being determined by the severity of the endometritis which has been developed. With the passage of time, and particularly of successive oestrous periods, the metritis should resolve spontaneously but local antiseptic or antibiotic therapy may be required in severe cases.

At service, some Caslicked mares will require to be partially opened to allow service especially those of group III and this should be done immediately before service and the wound resutured immediately afterwards. Opening may be unnecessary in dry mares Caslicked the previous season if care is taken to lubricate the vulval lips well with petroleum or KY jelly. At parturition, similarly, the Caslick repair may need to be incised prior to delivery of the foal's head, particularly in mares which have been operated upon repeatedly and in which fibrous tissue reaction has reduced the normal elasticity of the vulva. Again, the incision must be repaired immediately after delivery of fetal membranes. Once a mare has undergone a Caslick's operation for any reason, she almost always needs to remain in this state for the rest of her life.

The advantage of Pouret's operation is that no comparable interference is required.

C. EPISIOTOMY

I INDICATIONS

Episiotomy is rarely performed in veterinary practice but it has its place as a preventive measure, particularly in bovine obstetrics.

The objective is to substitute a surgical wound which will heal by first intention for a bruised and lacerated perineum. The procedure is, therefore, indicated in animals in which the vestibule and/or vulva are small or are resistant to stretching at delivery and this smallness is the sole cause of dystocia. The most likely causes of this constriction are an oversize fetus in a heifer, the presence of a particularly small vestibulo-vaginal junction, and a dry parturition with excessive friction.

Nevertheless, as Roberts (1971) points out, gentleness, patience and lubrication are often successful in such cases without the need for episiotomy.

(a) Technique

The procedure has been described by Freiermuth (1948) who states that he had employed the operation many times in the previous ten years. Swarbrick (1964) also advocated its use after employing it successfully in the U.K. for several years.

Anaesthesia may be by caudal epidural anaesthesia or, alternatively, the incision may be made without anaesthesia as the cow strains its hardest (experience in humans suggests this procedure is virtually painless when done at the time of maximum push).

As the calf progressively dilates the vulva, Swarbrick states that it is possible to determine by careful observation backed by experience when further traction will result in tearing - the vulva and surrounding tissues stand out clearly from the remainder of the perineum. An incision is made extending from about two thirds up the vulval lips on one side (the uppermost if the cow is recumbent), cranially and dorso-laterally to pass dorsal to the tuber ischii and lateral to the rectum (the anal sphincter must be avoided). Initially the incision is made only through the skin but if delivery is still impossible, it may be deepened to include the vestibular mucosa, perhaps as far cranially as the vestibular-vaginal junction. The length and depth of the incision should be enough to allow delivery without tearing but not so deep as to section the terminal branches of the vaginal artery. The wound is repaired after delivery. Arteries are clamped with haemostats and then ligated. Gross contamination is washed away. The mucous membrane is first sutured with absorbable material using a continuous everting pattern. A further layer closes the deadspace in the sub-cutaneous tissue and horizontal sutures are placed in the skin. It is important to produce as little distortion as possible. The investment of time and care is rewarded by rapid and satisfactory healing.

(b) Sequelae

Abscesses may develop. These must be lanced and drained as described on page 190. Necrosis of the vaginal wall may ensue. This may be prevented by topical application of antiseptic ointment to the vestibular mucosa for several days post-operatively. Occasionally the necrosis may be so severe as to cause rupture of the vaginal artery, usually with fatal consequences.

Swarbrick (1964) reports that future fertility is usually unimpaired and that there may be mild cicatrisation of the vulval lips.

D. PERSISTENT HYMEN

I ANATOMY

The vagina and vestibule of the female develop from two distinct embryological structures. The vaginal lumen develops by cavitation in the paramesonephric (Muellerian) duct whereas the vestibule develops by separation of the fetal cloaca into the reproductive and alimentary compartments.

Normally, the vaginal and vestibular lumina become continuous with one another because the membrane which separates them breaks down.

Occasionally, however, the membrane persists. As it is cranial to the urethral orifice, urination is not interfered with but, once the animal starts having ovarian cycles, mucoid uterine secretions build up behind the membrane. This condition, which the author has seen only in mares, must not be confused with segmental aplasia of the paramesonephric duct in cattle ('White Heifer Disease') or with a congenital tightness of the vulva and vestibule in animals which have never given birth.

II CLINICAL PICTURE

The mare may be presented because the owner has observed something protruding from the lips of the vulva. This something is the bulging hymen with uterine secretions behind it, and usually appears glistening white. Alternatively, it may be reported that the stallion is unable to serve the mare and vaginal examination reveals an obstruction just cranial to the urethral orifice. Such an obstruction may be distinguished from segmental aplasia of the paramesonephric duct by rectal examination - in persistent hymen the vagina and uterus will be distended. Colic has also been reported as the major presenting sign.

III TREATMENT

Persistent hymena may be readily ruptured using a blunt instrument. The hole should then be enlarged by the fist. Haemorrhage is rarely a problem.

E. CYSTS OF GAERTNER'S CANALS

I ANATOMY

The two canals of Gaertner are the remnants in the female of the fetal mesonephric (Wolffian) duct. When present in the adult, they run in the ventral wall of the vagina between the muscular and mucous coats. They may be up to ½ cm in diameter and may open caudally near the external urethral orifice. However, the duct may not be continuously patent throughout its length.

II PATHOLOGY AND CLINICAL PICTURE

The ducts may become cystic in cows poisoned with highly chlorinated naphtalenes or in cows with ovarian follicular cysts - in such cases, of course, cysts are not the most prominent clinical finding. Occasionally the ducts become cystic after an acute vaginitis.

Either a single spherical cyst or a row of cysts may develop, or a whole length of canal may become cystic taking a tortuous course. Occasionally the swelling can be as large as a cricket ball.

III SURGERY

The swellings should, in theory, be distinguished from abscesses or haematomas in the vaginal wall, but in practice abscesses, mature haematomas and cysts are all punctured and drained, and allowed to granulate.

F. CYSTS OF THE VESTIBULAR GLANDS

I ANATOMY

There are two collections of glands in the vestibule. The smaller glands lie well ventrally and open into the median ventral groove. The larger glands, often called Bartholin's glands, lie in the lateral walls under the vulvar constrictor muscles and are each about 3 cm long and 1.5 cm wide.

In the mare the ducts of the larger glands open on a group of 8-10 large prominences on the dorso-lateral wall of the vestibule. In the cow they open into a small pouch of mucous membrane which, in turn, opens onto the floor of the vestibule about 4.5 cm caudal to the urethral orifice and lateral to it.

II PATHOLOGY

These glands are sensitive to oestrogens and respond to high oestrogen levels by producing thin mucus and by developing hyperplasia of the duct epithelium. Cysts may, therefore, form in such conditions as ovarian follicular cysts. Alternatively, cysts may form because the ducts become blocked with inflammatory products from a vaginitis.

III SURGERY

The cystic swelling may be confused with vulval haematoma or abscesses but all three conditions are amenable to puncture, drainage and such irrigation as seems necessary.

G. VAGINAL VARICOSITY

Intermittent vulval bleeding may occur occasionally in mares with prominent 'varicose' veins at the vestibular/vaginal junction. On some of these distended veins, prominent bulges are apparent in which the outer walls of the vein appears to have ruptured and through which the inner wall is bulging these appear bright red rather than blue. The condition is commonest in old, multiparous mares (see White, Gerring, Jackson and Noakes, 1984).

In the cases seen by the author no treatment has been undertaken and one case has foaled since without any complications. The risk of fatal heomorrhage from such a lesion is like that from human haemorrhoids, extremely low.

H. HAEMORRHAGE IN THE MESOMETRIUM

In late pregnancy, the fragility of the utero-ovarian and the middle uterine artery is increased and there is increased tension on the broad ligament. Occasionally, this leads to rupture of the artery with either haematoma formation or massive internal haemorrhage and death. The condition occurs almost always at foaling and the prognosis for either condition is gaurded.

Haematoma may result in constipation and straining, as may abscesses which may develop several weeks after foaling. The may may lose condition. Manual exploration of the pelvic area may be restricted, but the swelling may only be detected at routine examinations as a cause of infertility. Surgical drainage through an incision in the vaginal wall is possible, but is unlikely to be successful if the lesion extends forwards to the ovary. If haemorrhage occurs it is difficult, if not impossible, to control.

J. REFERENCES

Arthur, G.H. (1975) Veterinary Reproduction and Obstetrics 4th Edition, p. 474. Bailliere and Tindall, London.

Blue, M.G. (1985). A tubo-ovarian cyst, paraovarian cysts and lesions of the oviduct in the mare.

Brown, M.P., Colahan, P.T. and Hawkins, D. (1978) Urethral extension for treatment of urine pooling in mares. Journal of the American Veterinary Medical Association 173 : 1005-1007.

Caslick, E.A. (1937) The vulva and the vulvo-vaginal orifice and its relation to the genital health of the thoroughbred mare. Cornell Veterinarian 27 : 178-187.

Freiermuth, G.J. (1948) Episiotomy in veterinary obstetrics. Journal of the American Veterinary Medical Association 113 : 231-232.

Monin, T. (1972) Vaginoplasty : a surgical treatment for urine pooling in the mare. Proceedings of the Annual Convention of the American Association of Equine Practitioners 18 : 99-102.

Pascoe, R.R. (1979) Observations on the length and angle of declination of the vulva and its relation to fertility in the mare. Journal of Reproduction and Fertility, Supplement 27, pp. 299-305.

Pouret, E.J.M. (1982) Surgical technique for the correction of pneumo- and urovagina. Equine Veterinary Journal 14 : 249-250.

Roberts, S.J. (1971) Veterinary Obstetrics and Genital Diseases. Published by the author, Ithaca, New York.

Rupp, G.P. and Kimberling, C.V. (1982) A new approach for spaying heifers. Veterinary Medical Small Animal Clinician 77 : 561-565.

Stabenfelt, G.H., Hughes, J.P. Kennedy, P.C., Meagher, D.M. and Neely, D.P. (1979) Clinical findings, pathological changes and endocrinological secretory patterns in mares with ovarian tumours. Journal of Reproduction and Fertility, Supplement 27, 277-285.

Swarbrick, O. (1964) Bovine episiotomy with a note on the comparative medical technique. Veterinary Record 76 : 359-362.

Walker, D.F. and Vaughan, J.T. (1980) Bovine and Equine Urogenital Surgery. Lea and Febiger, Philadelphia.

White, R.A.S., Gerring, E.L., Jackson, P.G.D. and Noakes, D.E. (1984) Persistent vaginal haemorrhage in five mares caused by varicose veins of the vaginal wall. Veterinary Record 115 : 263-264.

APPENDIX I

A. RESTRAINT AND ANAESTHESIA OF HORSES FOR CASTRATION

(a) Local Anaesthesia in the Standing Position

The method is applicable to both broken and unbroken horses but it is much easier in horses that have been extensively handled.

(i) The simplest way to anaesthetise a horse for castration in the standing position is as follows:-

  (1) The animal's right flank is placed against a wall and a twitch applied to the animal's upper lip.
  Tranquillisation can produce problems of the horse lying on the operator. If the horse is awkward to handle it may be better to administer a general anaesthetic than struggle on in the standing position. The lifting of the fore leg may steady an otherwise awkward animal.

  (2) The scrotum, the inner aspect of the thighs and the penis, prepuce and sheath are all washed with an antiseptic solution.

  (3) The operator stands with his left shoulder against the left side of the animal's chest and gently grasps the neck of the scrotum.
  The position adopted by the operator must be such that the possibility of his being injured in anyway by a cow kick from the horse is minimised. The scrotal neck is grasped gently and the testis milked into the scrotum gently as this disturbs the horse least. Tensing the testis into the scrotum too tightly may also make the horse object.

  (4) 10-20 ml of 3% Lignocaine are injected through a 19 gauge 4 cm long needle deep into the substance of the testis and as the needle is withdrawn, 2 ml of solution is placed immediately sub-cutaneously.
  This technique produces good analgesia of the spermatic cord, presumably by diffusion out of the veins of the pampiniform plexus, and also appears to produce good analgesia of the whole of the scrotal skin although the mechanism by which it does so is obscure. The horse does not usually object to the insertion of a fine gauge needle through scrotal skin into the substance of the testis, nor to the injection of anaesthetic solution into the testis.

  (5) The procedure in (3) and (4) is repeated on the opposite testis and the horse left for not less than 10 min and preferably not less than 20 min before surgery is commenced.

  There are two other ways of anaesthetising the spermatic cord and scrotum and these are as follows:-

(ii) In the nervous horse an insensitive skin weal is first made in the centre of the most dependant part of the scrotum with a ½" 25 or 23 gauge needle, and 2 ml of 3% lignocaine. A line from pole to pole of the testis is anaesthetised with about 10 ml of 3% lignocaine, using a longer (6 cm) needle inserted, if necessary, through the insensitive skin weal. According to Goulden (Personal Communication), it is important that the needle be parallel to the skin and immediately below it -if it enters the tunica dartos or testicle itself, the skin will not be desensitised. The procedure is repeated on the left scrotal skin. One minute is allowed to elapse before proceeding to anaesthetise the cord and testis by injection of 5-10 ml of 3% lignocaine into the cord via a long (not less than 6 cm long) needle passed through the substance of the testis into the cord.

  This technique has the disadvantage that a large gauge needle must be used and the horse may object to this being pushed through his testis.

(iii) The skin is anaesthetised as described above but the cord is anaesthetised by injecting 5-10 ml of 3% lignocaine through the neck of the scrotum.

  This technique is the least easy of the three as the horse has a very short scrotal neck. The technique is more readily applied to the ruminants - see Appendix II.

The *disadvantages* of the technique of standing castrations are the risk to the surgeon from a fully conscious horse and the fact that since the animal as a whole is neither anaesthetised nor completely restrained, any complication arising during the operation cannot be dealt with. Moreover, the level of surgical cleanliness obtainable is low. Its advantage is that the horse remains on its feet, so none of the risks associated with casting by ropes or anaesthesia are present. It is a method which impresses some horse owners but the author would neither use nor recommend it.

(b) Local Anaesthesia in the Cast Position

The horse is tranquillised and then cast with ropes by the London method as outlined in Figures I.1 and I.2. Once the horse is cast, the testis and scrotum are anaesthetised by the techniques described above (b).

The *disadvantages* of the method are that casting requires a team of men who know exactly what to do when the horse is being cast and roped up, and, as with the standing castration, complications arising during the operation cannot easily be dealt with. Its *advantages* are that the horse can be allowed to its feet immediately after the operation, and, compared to the standing operation, there is no danger to the surgeon.

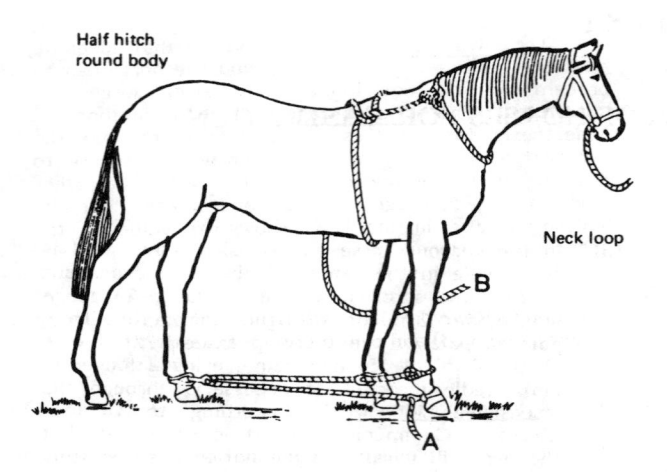

Fig. I:1 Horse prior to casting by the London Method with ropes placed in position

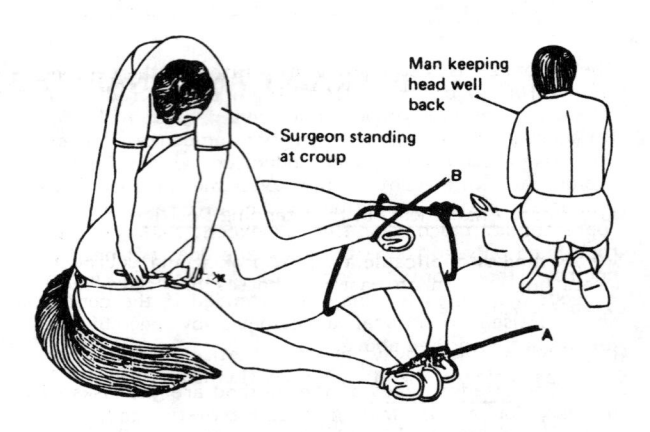

Fig. I:2 Horse cast and restrained with ropes for castration.

Two ropes each about 10 metres long are required together with three hobbles. Each rope should have a spliced noose at one end.

A loop is made at the end of Rope A with the spliced noose and placed above the metacarpo-phalangeal joint (fetlock) of the right fore leg. The rope is then threaded in succession through the loops of hobbles placed round the first phalanx (pastern) of the left fore leg, the left hind leg and the right fore leg. The rope is held on the animal's right hand side.

A neck look is made from Rope B and fixed as shown by a knot so that it cannot pull tight round the neck. A half hitch is placed round the thorax and the rope held on the animal's left hand side.

The head is held by a man whose job it will be to keep the head of the horse well back when the horse is on the ground - this is an extremely responsible position.

When all is ready, one man pulls to the horse's left on Rope B and two men (preferably) pull on Rope A to the horse's right, starting as close to the right fore leg as possible, and a fourth man controls the head.

What to do next is described under Fig. I.2.

When the horse has been pulled to the ground as described in Fig. I.1, the man on the head must keep the horse's head well back - failure to do so may result in the horse breaking its back.

Rope B is looped twice round the first phalanx (pastern) of the right hind leg (for clarity only one loop is shown) and then tied to the neck loop so as to restrain the right hind leg cranially and dorsally and to expose the inguinal region.

Rope A is used to secure the three hobbled legs by taking two half hitches round the first phalanx (pastern) of the right fore leg. It is necessary for Rope A to be held by a man throughout the surgical procedure.

The surgeon stands by the horse's croup and leans over the horse's back to castrate him. He operates on the distant testis first.

When the operation is completed, the right hind leg is released from Rope B, which is then itself left free. The half-hitches of Rope A round the right fore leg are released and Rope A allowed to run freely through the hobbles. Now, and only now, the man on the head may stand and encourage the horse (held only by a lead on the halter) to stand. Once the horse is standing, the ropes and hobbles may be removed.

NOTE:— These instructions refer to the position for a right-handed surgeon. For a left-handed surgeon, the horse should be cast on its right and the instructions should be altered accordingly.

(c) **Intravenous Administration of Short-Acting Anaesthetic**

The horse is tranquillized. Either thiopentone (1.1mg/100Kg) or thiamylal (0.74gm/100Kg) sodium is given as an intravenous injection. (Hutchins and Rawlinson (1972) recorded a higher incidence of post-operative herniations after induction with thiamylal sodium and the agent is not available in the U.K.) Immediately after induction the horse is restrained by ropes applied much as in Fig. I.2 except that the body rope need only go round the neck and does not need to encircle the thorax. Length of anaesthesia rarely exceeds 5 minutes but may be extended if the cord is anaesthetised with local anaesthetic by one of the techniques described above.

The disadvantages of the method are the risks of perivascular administration of anaesthetic and of a general anaesthetic itself, and the fact that the horse has an excited recovery phase. The latter means that the operation is preferably done in an open field. Excitement can, however, be minimised by covering the horse's head (but not his nostrils) with a blanket or sack during the recovery period. The advantages of the method over the use of local anaesthesia alone are that the horse is under the operator's complete control and anaesthesia can be prolonged if necessary by supplementary doses of barbiturate or by an inhalational agent.

A variant of the method is to use chloral hydrate (14gm/100Kg) instead of an ultra-short acting barbiturate. Chloral hydrate has the advantage over ultra-short acting barbiturates that it provides a more prolonged period of anaesthesia. Moreover, it allows the horse to be placed in dorsal recumbency without rope restraint allowing easy performance of a 'closed' operation (see p.12).

(d) **Administration of an Inhalational Agent**

In a veterinary practice with closed circuit anaesthesia, halothane and oxygen through an endotracheal tube can be used to prolong anaesthesia induced by a short acting barbiturate.

Alternatively, chloroform may be administered via a Cox's or McCunn's mask either to the standing horse or to the horse cast with ropes. The method is particularly suitable for pony colts (Lewis, 1960) or for wild or unbroken horses. 30-60ml of chloroform is poured onto the sponge placed in the mask. The standing horse should not be premedicated, as the more it 'fights' the chloroform the more quickly it succumbs. The particular advantage of the method is the speed of recovery, the recovery also being without much excitement.

(e) **'Immobilon/Revivon'**

'Immobilon', which contains etorphine hydrochloride (a potent morphine analogue) and acepromazine, is administered, preferably intravenously, at a dose not less than 1ml/100Kg, to cast and anaesthetise the horse and 'Revivon' which contains an antagonist to the etorphine, is administered post-operatively to bring the horse back to its feet. ('Immobilon/Revivon' are not available world wide). There is a tendency for the operator to want to move in and castrate the animal soon after it becomes cast, but the muscle tremors and general rigidity of the animal induced by etorphine tend to pass off after a few minutes, and a little patience is rewarded by a more relaxed patient.

The advantages of the method are that adequate anaesthetic time is provided and that the horse can be brought to its feet immediately after surgery is completed. The disadvantages of this method of anaesthesia for purposes of castration are that the horse's hind legs may move vigorously in response to crushing of the spermatic cord. The mixture has other inherent disadvantages as an anaesthetic but its widespread use in the U.K. shows its value to the general practitioner.

Rare complication of the use of 'Immobilon/Revivon' are intestinal prolapse during surgery (see p.63) and penile prolapse (see p.77).

(f) **Xylazine/Ketamine**

A recently introduced combination is xylazine (1.1mg/Kg given by slow intravenous injection) followed 2-3 minutes later by ketamine (2.2mg/Kg by rapid intravenous injection). The method produces excellent sedation, smooth induction and adequate surgical anaesthesia for an open castration. The combination is characterised by a quiet recovery, probably because of the excellent properties of the xylazine. It is, however, expensive.

REFERENCES

Hutchins, D.R. and Rawlinson, R.J. (1972) Eventration as a sequel to castration of the horse. Australian Veterinary Journal 48 : 288-291.

Lewis, D.G. (1960) Equine castration. Veterinary Record 72 : 212-213.

APPENDIX II

RESTRAINT AND ANAESTHESIA OF FARM ANIMALS FOR CASTRATION

(a) The calf may be restrained in any satisfactory way which exposes the scrotal area to the surgeon. The simplest way is for an assistant to hold the calf's head with one hand and the tail with the other whilst using his knee and body to hold the calf against a wall or partition. However, improved exposure of the scrotum (see Fig. II.1) is obtained by having the calf's head tied by a halter to a suitable post (or, in the older animal restrained in a yoke) whilst the assistant holds up the tail with one hand and holds one leg laterally by the hock as well as keeping the calf held firmly against the wall or partition.

Anaesthesia is satisfactorily produced by the injection of 5-20 ml of 3% lignocaine directly into the testis and waiting 5 minutes. This method gives adequate anaesthesia of the spermatic cord and testis and, provided sufficient time is allowed to elapse, gives anaesthesia of the scrotal skin (see also p.i). This technique should NOT be employed if a Burdizzo (see p.33) is to be employed because of the risk of introducing infection.

(b) The lamb is best restrained in a sitting position, as shown in Fig. II.2 as this gives good exposure of the scrotum and keeps the legs under control.

Anaesthesia is produced as in the calf, except that only 2-10 ml are needed, according to the size of the testis.

(c) The piglet. Piglets are easily restrained by holding each hind leg round the hock. Piglets up to three weeks of age can then be tucked under the arm of the assistant. Larger pigs may need to be restrained between the legs of the assistant.

Anaesthesia may be produced as described for the sheep.

(d) The boar. Occasional requests are made by farmers to have adult boars castrated (and de-tusked). (Not less than six weeks, after castration, the boar can be sent for slaughter without fear of taint being present in the meat). Such animals, however, pose considerable problems of restraint and anaesthesia.

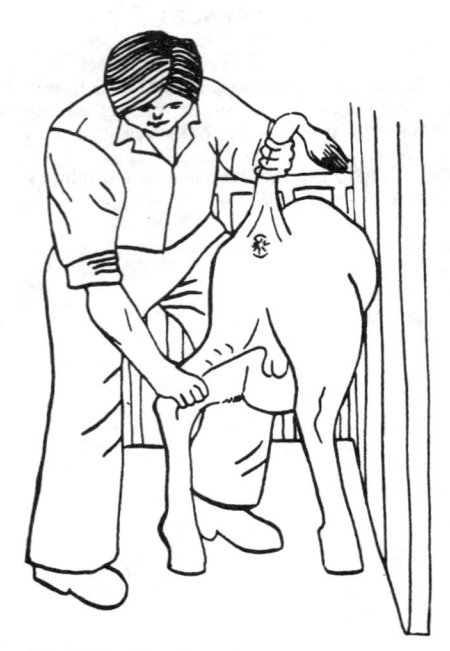

Fig. II:1 Showing assistant restraining calf for castration with the inguinal region well exposed. The calf's head is haltered and tied to a post.

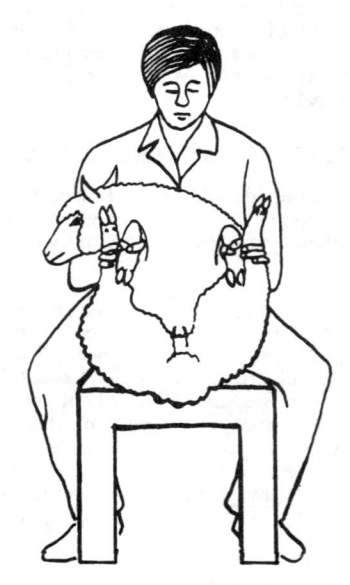

Fig. II:2 Showing assistant restraining a lamb for castration. Note that both fore and hind legs of the lamb are restrained. The assistant sits comfortably astride a bench.

(i) **Local Anaesthesia**

Local anaesthesia can be produced by any of the techniques described above for the horse (pp.i-ii). However, there are problems of restraint. One technique is to run the boar's head and chest into a dustbin and to up-end the dustbin, thus bringing the scrotum to hand level for surgery. Another technique is to restrain the boar's head with a nose snare tied securely to a strong post or ring. Neither technique is particularly humane and the noise produces temporary deafness in both surgeon and assistants. Moreover, these forms of restraint and anaesthesia apparently increase the shock of the operation and a proportion of boars so restrained will die within 24 hours. Although the boar may be tranquillised with azaperone, the dose required to produce immobility (8mg/Kg) involves injecting a large volume of solution and is not certain in its results. The induction of general anaesthesia is, therefore, preferable, although such a procedure always carries with it a higher risk in the pig than in any other species.

(ii) **Conventional General Anaesthesia**

Prior to induction of general anaesthesia, the boar must be starved of food and water for twelve hours. Although any number of methods of procuring general anaesthesia are possible, the most suitable in terms of safety and speed of recovery is to tranquillize the boar with azaperone at a dose rate of 4mg/Kg by deep intramuscular injection 15 minutes before induction with 'Saffan' given intravenously at a dose rate of 2-3mg/Kg. An advantage of premedication with azaperone is that the ear veins become dilated and intravenous injection of 'Saffan' is facilitated. The advantages of 'Saffan' over metomidate are that it provides anaesthesia without the need to supplement with local anaesthetic and that it is followed by a faster recovery.

(iii) **Intra-testicular Administration of General Anaesthetic**

This method of administering anaesthetics solves the problem of finding a suitable ear vein and of injecting anaesthetic into it. It is, of course, just as necessary to starve the boar for 12 hours as with conventional anaesthesia.

In order to retrain the boar whilst the injection into the testis is made, the use of a nose snare is necessary. Surprisingly, the injection into the testis does not seem to provoke much pain. Two systems have been described.

The *first* system involves the use of pentobarbitone sodium solution at a strength of 200 mg/ml - this is the solution usually used for euthanasia. A dose is calculated for the animal on the basis of 45-50mg/Kg and half this dose is injected into each testis. The boar generally lies down in 5 minutes and full surgical anaesthesia is reached in a further 5 minutes. Recovery takes 20-40 minutes (Dyson, 1965).

The *second* system is to use azaperone and metomidate. Nitz (1974) gave 400 mg of azaperone and 1 gm of metomidate to all boars over a weight range of 100-300 Kg. The boars became recumbent after 4 minutes and surgical anaesthesia was reached in 6 minutes. Recovery, as might be expected following the use of metomidate, was prolonged, being 2-3 hours.

Care must be taken with either system that the anaesthetic agent(s) goes into the testis and does not leak into the lumen of the vaginal process. Moreover, the testes must be disposed of in such a way as to prevent dogs or other pigs from eating them.

REFERENCES

Dyson, J.A. (1964) Castration of the mature boar with reference to general anaesthesia induced by intratesticular injection of pentobarbitone sodium. Veterinary Record 76 : 28-29.

Nitz, K.J. (1974) Eberkastration - unproblematisch und einfasch. Der Praktische Tierartzliche 6 : 323-324.

APPENDIX III

PUDENDAL NERVE BLOCK IN THE BULL

In order to operate upon the penis and preputial membrane of the bull, it is necessary to have both analgesia of the integumentary surfaces and relaxation of the retractor penis muscle. Whilst both objectives may be achieved be general anaesthesia, this procedure carries with it a risk of complications, for example, regurgitation and inhalation of rumen contents. For most simple operations or for examination of the penis and adnexa this risk is high.

Anaesthesia can be produced by epidural block but in practice it is not always possible to achieve satisfactory anaesthesia and relaxation of the penis without also affecting the bulls ability to stand. As there are management problems in bulls with hind limb inco-ordination, the technique of epidural block for penile anaesthesia is also unacceptable.

Pudendal Nerve Block, has few of the disadvantages outlined above. The bull may be kept standing, or he may be cast if that is considered desirable. Although a number of approaches have been described, the most satisfactory one is that through the ischio-rectal fossa as first described by Larsen (1953). The technique described hereafter has been in use at the Liverpool Veterinary School for many years.

Figures III.1 and III.2 should be studied independently with their relevant text. They may then be superimposed to study the relationship between them. The section on 'Clinically Palpable Structures' is best studied with a living bovine animal which may be palpated both externally and internally.

Pudendal Nerve Block in the bull sounds much more complicated to perform than it actually is. The sites for the block are readily palpated in the thin dairy bull, in which initial experience of the technique should be sought. There is difficulty in inserting a long needle towards points which cannot always be palpated and which have no externally palpable landmarks. This difficulty is compounded by a fear of penetrating the rectal wall. Once these initial doubts have been overcome the technique is quickly mastered and will be found to have a wide application.

A. ANATOMICAL CONSIDERATIONS

I. THE NERVES OF THE SACRAL PLEXUS (Figure III.1)

A. Caudal Rectal Nerves

Origin = S4 and S5, the latter being generally small.

Supply amongst other things, motor function to the proximal segment of the retractor penis muscle.

Blocked by local anaesthetic placed at B. Probably not necessary for penile relaxation.

B. Pudendal Nerve

Origin = S3 with contributions from S4 and (not shown) S2, and possibly also S1 and L6.

Branches

1. Proximal Cutaneous Nerve
2. Distal Cutaneous Nerve
3. Deep Perineal Nerve

(These branches are not significant in pudendal nerve block for penile relaxation and anaesthesia).

The Pudendal Nerve Proper then splits into

(i) Dorsal Nerve of Penis, which supplies motor function to the distal segments of the retractor penis muscle, and sensation to the free end of the penis.

(ii) Superficial Perineal Nerve which in turn splits into

(a) Preputial branch which supplies sensation to the preputial membrane.

(b) Scrotal branch, which in part supplies cutaneous sensation to the scrotum.

The whole of the pudendal nerve is blocked by local anaesthetic placed at A.

C. Pudendal Branch of the Ischiatic Nerve

(= Medial Branch of Caudal Cutaneous Femoral Nerve)

The ischiatic nerve (origin S2, S1 and L6) leaves the pelvis through the greater ischiatic foramen, thus coming to lie lateral to the broad sacro-tuberous ligament. Soon afterward emerging from the foramen it gives off the caudal cutaneous femoral nerve which is directed caudally. This nerve has a medial branch which re-enters the pelvis through the lesser ischiatic foramen to join the pudendal nerve proper.

Blocked by local anaesthetic placed at C.

NOTE: There is marked individual variation in branching and anastomosis of these nerves.

(vii)

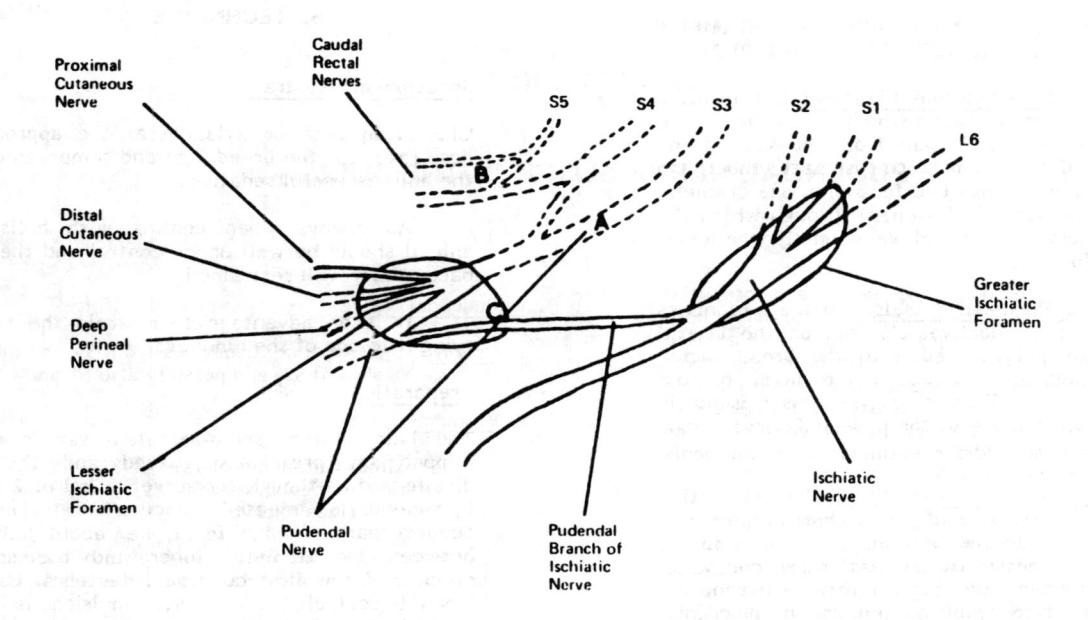

Fig. III:1 Nerves of the sacral plexus

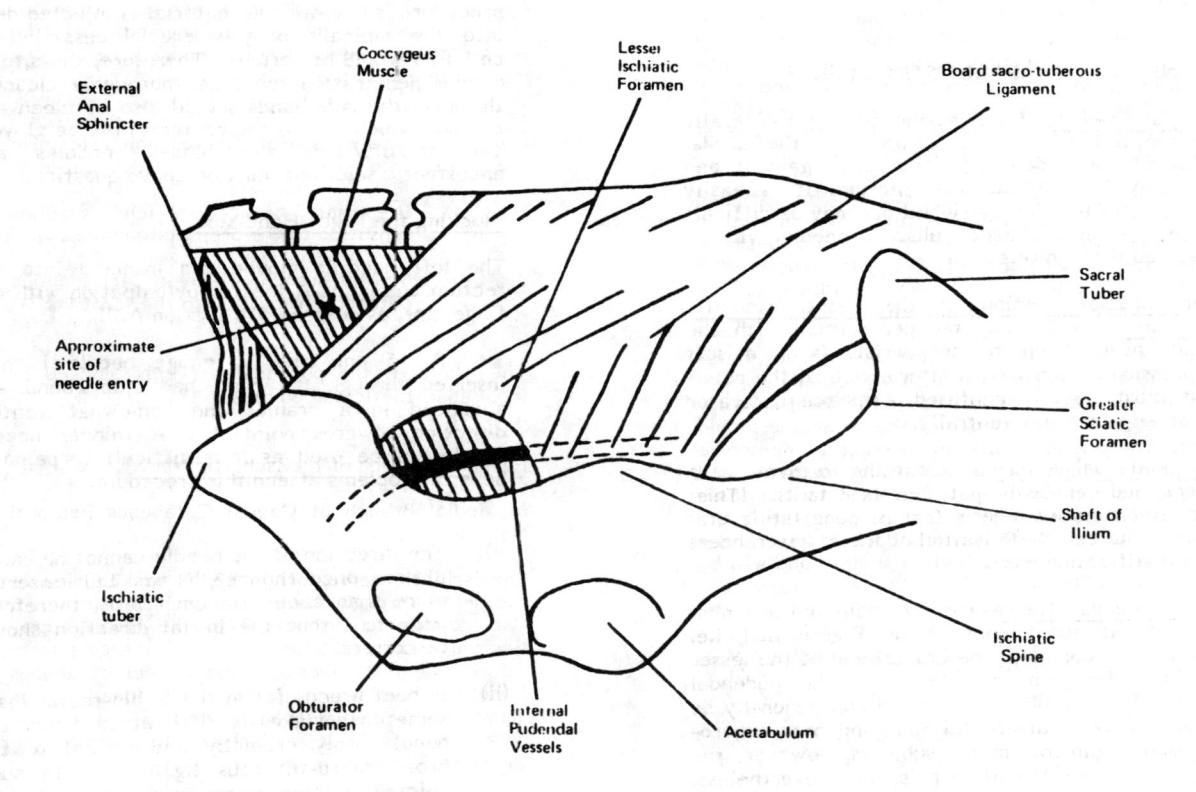

Fig. III:2 Broad sacro-tuberous ligament and related structures: Lateral view with lateral muscles removed

II THE BROAD SACRO-TUBEROUS LIGAMENT AND RELATED STRUCTURES (FIGURE III.2)

A. The Broad Sacro-tuberous Ligament is a massive sheet of connective tissue which forms the lateral wall of the pelvis. It runs from the sacrum and the shaft of the ilium to the ischiatic tuber and ischiatic spine. It has two foramina, one cranially (the greater ischiatic foramen) through which the ischiatic nerve passes and one caudally (the lesser ischiatic foramen).

B. The Internal Pudendal Vessels, which are branches of the internal iliac vessels, run on the lateral wall of the pelvis, medial to the broad sacro-tuberous ligament at about the level of the ischiatic spine. They cross the lesser ischiatic foramen to run close to the pudendal nerve at the caudal end of the foramen.

C. The Coccygeus Muscles are sheet-like, originating from the medial side of the ischiatic spine and passing lateral to the rectum to insert on to the transverse processes of the first three coccygeal vertebrae. When palpating per rectum the muscle will be found to cover all but the most cranial portion of the lesser ischiatic foramen. Although the caudal rectal nerves pass medial to the muscle, the pudendal nerve and its branches pass lateral to it.

III CLINICALLY PALPABLE STRUCTURES

A. The ischio-rectal fossa is bounded ventro-laterally by the ischiatic tuber, cranially by the caudal edge of the broad sacro-tuberous ligament and medially by the tail and rectum. It is easily palpable in the thin dairy bull but may be difficult to define in fat beef bulls. A needle will be inserted through this area.

B. The Cranial Portion of the Lesser Ischiatic Foramen may be palpated per rectum with the hand inserted up to the wrist. It is a soft depression on the ventro-lateral wall of the pelvis and must not be confused with the obturator foramen which lies ventrally.

C. The Pudendal Artery is about the thickness of a pencil and can be palpated cranial to the lesser ischiatic foramen where it runs along the lateral wall of the pelvis. Pulsation of the artery renders its identification easy.

D. The Pudendal Nerve can be palpated in most subjects at about point A on Figure III.1 i.e. cranial and dorsal to the cranial end of the lesser ischiatic foramen and dorsal to the pudendal artery. It feels like a straw and can generally be rolled between the rectal wall and broad sacro-tuberous ligament. In fat subjects, however, the nerve may be difficult to palpate. Nevertheless, point A can always be defined by its relationship to the pudendal artery (which is palpable ventrally) and to the cranial portion of the lesser ischiatic foramen (which is palpable ventro-laterally).

NOTE: The caudal rectal nerves and the pudendal branch of the ischiatic nerve cannot be palpated.

B. TECHNIQUE

1. Sedation and Restraint

Chloral hydrate or xylazine at the appropriate dose rates for the breed, age and temperament of the bull are useful sedatives.

As always when dealing with bulls, the animal should be well under control and the head particularly well restrained.

It is an advantage to restrain the tail by tying it to one of the hind legs.

2. Preparation

The hair in the ischio-rectal fossae is close-clipped or, preferably shaved and the skin disinfected as though for surgery. 2ml of 2 or 3% lignocaine is injected subcutaneously in the deepest part of each fossa, i.e. about half-way between the ischiatic tuber and the spinous process of the first coccygeal vertebra. Using a No. 11 scalpel blade a stab-incision is made through the skin in this area - this will facilitate insertion of the long needle.

Rigid surgical cleanliness throughout the procedure is essential as material is injected deep into the aminal's body where abscessation or cellulitis would be serious. Therefore, the site of needle penetration must be thoroughly cleaned, the anaesthetist's hands should also be clean and the one which lies in the rectum must be gloved. The sterility of the blades, needles and anaesthetic solutions must be above question.

3. Blocking the Right-Hand Side

The left hand is gloved and inserted into the rectum and point A located by palpation with the fingertips, as described in section A.III.

A 15 cm long 14 gauge needle is now inserted through the right had stab-wound and advanced in a cranial and somewhat ventral direction towards point A. A thinner needle should not be used as it is difficult to palpate. Several problems attend this procedure:

(i) The direction of the needle cannot be easily altered once the needle has been inserted more than about 4-5 cm - it is, therefore, important that the initial direction should be correct.

(ii) In beef breeds fat in the ischio-rectal fossa sometimes makes it difficult to direct the needly truly cranially and medial to the broad sacro-tuberous ligament. In such animals it is necessary to insert the needle very close to the fat base of the tail.

(iii) The needle cannot be palpated for the first 6-7 cm because the coccygeus muscle lies between it and the rectal wall.

(iv) There is, therefore, a fear of penetrating the rectal wall. This difficulty is best overcome by deflecting the rectum to the left side until the needle is judged to be in the correct position and then confirming this by palpation.

Whilst the anaesthetist holds the needle firmly in place, an assistant injects 8-10ml of anaesthetic solution (2 or 3% lignocaine hydrochloride) to block the pudendal nerve at A.

The needle must now be almost completely withdrawn and redirected somewhat dorsally towards point B. Palpation will again confirm that it is in the correct position. 5ml of anaesthetic solution are injected by an assistant to block the caudal rectal nerves.

Once again the needle is almost withdrawn and then redirected more ventrally to point C at the cranial part of the lesser ischiatic foramen. After palpation has confirmed that the needle is correctly placed a further 5ml of anaesthetic solution are injected to block the pudendal branch of the ischiatic nerve. As there is some risk of accidental injection of the pudendal vessels it is a wise precaution to have the assistant draw back on the syringe before injection. Whilst it is possible that local anaesthetic injected at C will block the pudendal nerve itself, it is not wise to inject large volumes in this region because of the proximity of the blood vessels.

The needle is now completely withdrawn and the area around points A, B and C gently massaged to disperse the anaesthetic solution and thus render slight inaccuracies in anaesthetic placement insignificant. Care, however, should be taken not to direct anaesthetic too far cranial to position A, or blockage of part of the ischiatic nerve may occur.

4. Blocking the Left-Hand Side

The procedure described above is now repeated for the left-hand side. The author finds it easier to strip off the left glove, to glove the right hand and place it in the rectum, and to insert the needle with the left hand through the left hand stab-wound. It is preferable to use a new sterile needle for the second side.

5. Effects of the Block

In the normal animal, prolapse of the penis may be evident immediately after anaesthetic placement or massage, but it is not usually delayed by more than ten minutes. The use of lignocaine is preferable to procaine, and massage of the area both appear to spread the onset of the block.

Anaesthesia and exteriorisation of the penis are adequate for detailed examination of the penis and preputial lining and for surgical or other treatment of these structures to be carried out.

It is a rule of pudendal nerve block that if the free end of the penis is desensitised, then motor paralysis of the rectractor penis muscle is present. (Some believe this is because retractor muscle tone depends on afferent impulses from the penis and prepuce). This rule has a useful diagnostic application. Failure to exteriorize the penis in the presence of analgesia of the free end cannot be due to continued normal tone in the retractor penis but must be due to some other physical impediment such as retractor muscle myopathy, preputial or peripenile adhesions or a congenitally short penis.

6. Post-anaesthetic Care

Once investigation or treatment has been completed, the penis should be lubricated with antiseptic cream and replaced in the sheath. Application of a bandage round the sheath for about four hours retains the penis in position until the block wears off, so preventing both drying from exposure and trauma by the patient.

REFERENCES

Larsen, L.L. (1953) The internal pudendal (pudic) nerveblock for anaesthesia of the penis and relaxation of the retractor penis muscle. Journal of the American Veterinary Medical Association 123 : 18-27.

APPENDIX IV

PARAVERTEBRAL ANAESTHESIA IN CATTLE

For Caesarean operation, the 13th thoracic and 1st, 2nd and 3rd lumbar nerves are anaesthetised.

The technique for locating the nerves which is given below is that used for many years at the Liverpool Veterinary School and gives good results even in the hands of the inexperienced.

1. <u>The 2nd lumbar nerve</u>

The 6th lumbar transverse process is medial to the external angle of the ilium, so the first palpable transverse process in front of the tuber coxae is the 5th (Fig. IV.1). The lateral extremity of the 2nd transverse process is located by counting forward from the 5th. The curvature of this transverse process is outwards and <u>forwards</u> so that the caudal edge is slightly behind the lateral extremity.

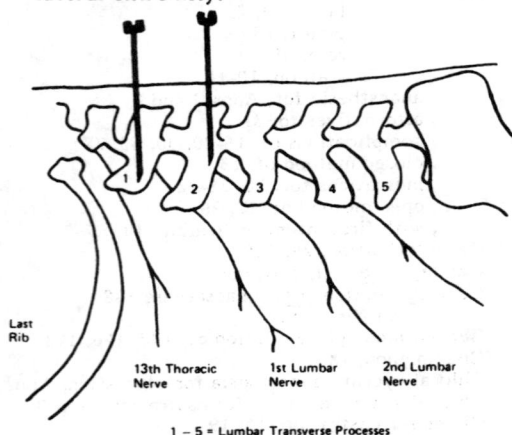

Fig. IV:1 Showing the ventral branches of the 13th thoracic and the lumbar nerves and their relationship to the transverse processes.

A small volume of 2% lignocaine without adrenaline is placed under the skin over the mid point of the second transverse process. A 4" 16 gauge needle is introduced until it strikes the transverse process, withdrawn slightly and then redirected off the caudal edge until it penetrates the inter-transverse ligament (Fig. IV.2). Failure to penetrate the ligament will result in no anaesthesia of the ventral branches and these are the more important. It is very helpful to strike the bone with the needle as one then knows just how deep the transverse process is and, therefore, how deep the inter-transverse process ligament lies. 12-14ml of 3% lignocaine are therefore, injected below the ligament in order to block the ventral branch. The remaining 6-8 ccs are injected as the needle is withdrawn thus blocking the dorsal branch of the nerve.

3. <u>The 1st lumbar and 13th thoracic nerves</u>

The lateral extremity of the 1st lumbar transverse process may not be palpable but its position can be calculated as it is as far in front of the second as the third is behind.

A small volume of 2% lignocaine without adrenaline is placed under the skin over the mid point of the transverse process approximately 2" from the mid-line.

The 4" needle is then introduced until it strikes the process, withdrawn slightly and then redirected first off the caudal edge to block the 1st lumbar nerve and then the cranial edge to block the 13th thoracic nerve.

In each case 12-14ml of 3% lignocaine are injected in order to block the ventral branch. The remaining 6-8ml are injected as the needle is withdrawn.

3. <u>The 3rd lumbar nerve</u>

This is anaesthetised in a similar fashion to the second.

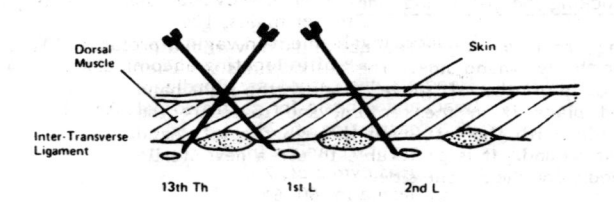

Fig. IV:2 Showing the relationship between the ventral branches of the spinal nerves and the inter-transverse ligament

INDEX

A

Aanes operation for recto vaginal tear, 173
Abdominal hernia - see Rupture
Abscess in peripenile tissues, 107
    vagina, vestibule, vulva, 172, 191
    following castration, 18, 19
    following penile haematoma, 104, 105, 107
Abortion as an alternative to caesarean, 149
    as a sequel to caesarean, 167
Acquired hernias, 61
Adhesions after caesarean, 166
    penile trauma, 95, 105, 107, 108
    preputial trauma, 95, 112
Afterbirth - see fetal membranes
Amputation of bovine penis, 119
    equine penis, 84
Anaesthesia for castration, Apps. I & II
    caesarean, 151
    other procedures - see procedures
    , paravertebral, App. IV
    , pudendal, App. III
Anorchid, 37
Antibiotic as alternative to caesarean, 148
Antibiotic pre-operatively, 12, 153
Apical ligament of bovine penis, 90, 100-101

B

Balanitis - see Balanoposthitis
Balanoposthitis in bull, 95, 102, 106, 107
    horse, 77
    ram, 107
Bartholin's glands - see Vestibular glands
Behaviour of cryptorchids, 42
    false rigs, 43
    mares with ovarian abnormalities, 183
    teaser males, 124
Bladder, involvement in vaginal prolapse, 131, 136
    , use of in locating abdominal testes, 47
Bloodless castrator - see Burdizzo
Boar, anaesthesia for castration of, App. II
    , castration of, 29
    , cryptorchidism in, 37
    , epididymis of, 7
    , hernia in, 60, 65
    , vasectomy in, 119
    - see also piglet
Bull, absence of libido in, 94
    , balanoposthitis in, 95, 102, 106, 107
    , castration of, 29
    , copulatory inability in, 94-95
    , cryptorchidism in, 37
    , hernias in, 62, 65
    , pudendal nerve block in, 96, App. III
    , vasectomy in, 119
Burdizzo method of castration, 24

C

Caesarean operation, 145 - 170
    , alternatives to, 147
    , anaesthesia for, 151, App. IV
    , antibiotics in, 153, 160
    , cervical section versus, 148
    , choice of incision site, 154
    , delivery versus, 147
    , fertility following, 167
    , fetal membranes after 158, 165
    , fetotomy versus, 147
    , incising uterus, 156
    , indications for, 145
    , maternal recover rate, 146, 167
    , monsters in, 180
    , operative procedure for, 154
    , post-operative course, 163ff
    , removal of fetus in, 158
    , slaughter versus, 150
    , suturing uterus in, 158
    , symphysiotomy versus, 148
    , wound repair in, 159
Calf, castration of, anaesthesia for, App. II
    , complications with, 30
    , restraint for, App. II
    , using Burdizzo, 24-26
    , using knife, 26-29
    , using rubber rings, 23-24
Canals of bovine penis, 87, 114
Cancer - see Tumour
Caslick's operation for pneumovagina, 188
Casting for castration, App. I
Castration by Burdizzo, 24-26
    rubber rings, 23-24
    of boar, 29
    bull, 29
    calf, 23-29
    colt, 10-15
    donkey, 15
    lamb, 23, 29
    piglet, 29
    ram, 29
    stallion, 10-15
    , anaesthesia for, Apps. I and II
    , alternatives to, 31
    , complications of, 16-20, 30, 63, 78
    , closed method of, 13
    , indications for, 10, 21-22
    , open method of, 12, 26
    , with first intention healing, 14
Catheterization, 74, 189
Cattle, - see bull, calf, cow
Cervical section versus caesarean, 148
    surgery, 181
Cervix, incomplete dilation of, 132, 146, 148
Champignon, 18
Chloral hydrate anaesthesia for castration, App. I
Chloroform anaesthesia for castration, App. I
Clams for castration, 15, 19
Clitoris, tumours of, 81
Coital exanthema, 75
Coitus, defective, 94-96
    - see also Erection
Colt, castration of, 10-15
    , hernia in, 60, 61, 62
Congenital abnormalities - see specific structure
Congenital hernia, 59
Corkscrew deviation of bovine penis, 101
Corpus cavernosum penis, 71, 87, 91, 104, 114, 124
Corpus spongiosum penis, 71, 87, 91
Cows, caesarean in - see caesarean operation
    , prolapse of vagina in, 127-143
    , removal of ovary from, 185
Cryptorchidism, definition of, 37
    , diagnosis of, 43
    , causes of, 41
    , induced as a teaser, 123
    , phenomenon of, 38
    , removal of single testis from, 43
    , significance of, 42
    , surgery of, 45
Cut-proud - see False rig
Cystic testicles, 32, 48
    ovaries, 183
Cysts in inguinal region, 81
    spermatic sac, 20, 33
    of Gaertner's canals, 192

vestibular glands, 192

D

Delivery per vaginam as alternative to caesarean, 147
Descent of testis, 1 - 5
Detumescence of bovine penis, 91
Deviations of penis in bull, 100-102, 121
Donkey, abscence of smegma in, 72,
    , castration of, 15
Dourine, 76
Dropsy of fetal membranes - see Hydrallantois
Dystocia, alternatives for, to caesarean, 147
    , prevention of, by caesarean, 146

E

Ecraseur, use of, in ovariectomy, 184
Emasculator, design of, 16
    , use of, in castration, 13, 14, 29, 48
Embryotomy - see Fetotomy
Emphysema, sub-cutaneous, after caesarean, 159, 166
Emphysematous calf, caesarean as a treatment for, 161
Epididymis, 1, 7, 8, 39, 40, 41, 46, 47
    injection of, for teaser, 123
Epididymectomy, 123
Epidural in caesarean, 152
    , versus pudendal nerve block, App. IV

Episiotomy, 150, 193
Epispadias, 99
Erection haemorrhages, 110
    , mechanism of in bull, 91
    stallion, 72
Ewe, caesarean in, 152, 154, 157, 159, 161, 164
    - see also caesarean
    , prolapse vagina in, 127-143
Examination of bull penis, 94-96
    horse for cryptorchidism,
    penis, 74
    prior to castration, 75, 11

F

False rig, 43
Fetal membranes following caesarean, 158, 165
Fetal monsters at caesarean, 160
Fetal oversize as indication for caesarean, 145
Fetotomy versus caesarean, 147
Fibropapilloma in bulls, 109
    horses, 81
Fistula following castration, 19, 74
    , recto-vestibular, 171, 174
Fluting, 186
Foal, castration of, 14
    , hernia in, 60, 67
Fracture of penis - see penile haematoma
Frenulum, persistence of penile, 98
Frostbite of scrotum, 33

G

Gaertner's canals, 192
Gelding behaving like a rig, 43
    , scrotal swelling in, 17, 20, 33
Granulosa cell tumour, 183
Gubernaculum, remnants in adult, 1, 8, 39, 46
    , role in testis descent, 2, 3, 4, 41
    , structure of, 1-3
Gut-tie, 30

H

Haemangioma of penis, 110
Haematocele, 33
Haematoma of penis, 77, 103, 106, 107
    vagina, vestibule and vulva, 172, 191
Haemorrhage after castration, 16, 30, 49
    parturition, 172
    in mesometrium, 193
    from uterus at caesarean, 153, 158
Heifer, pelvic symphysiotomy in, 148
Hernia, definitions, 53
    , femoral - see Rupture of Prepubic Tendon
    , pelvic - see Gut-tie
    - see also Inguinal hernia
Herniorraphy needle, 67
Horse pox, 75
Hydrallantois and caesarean operation, 162
    , other treatments, 149
Hydrocele, 33
Hydrocephalus, delivery of in sow, 147
Hydrops - see Hydrallantois
Hymen, persistence of, 191
Hypodermis of bovine penis, 92 - 93
Hypospadias, 99
Hysterotomy - see caesarean

I

Impotence, 94, 104, 114
Inability to serve, 94 - 96
Infantile penis, 97
Inguinal canal, 2, 3, 33, 54 - 56
    hernia, acquired, 61
    , anatomy of, 53, 59
    , congenital, 59
    , definition of, 53
    in bull, 60, 62, 65
    foal, 60, 67
    pig, 60, 65
    ram, 60, 62, 65
    stallion, 60, 62, 67
    , repair of, 67 - 69
    rings, 38, 45, 54 - 56
    - see also Vaginal ring
Intestines, prolapse of after castration 17, 30, 63
    through vagina, 178

L

Lacerations of penis and prepuce, 77, 107, 111
    vagina, vestibule and vulva, 171, 172
Laparotomy for caesarean, 154
    cryptorchidectomy, 47
    ovariectomy, 183
    post-castration bowel prolapse, 68
Legislation governing castration of
    farm animals, 21 - 23
    horses, 10
Libido, 94
Ligatures, use of, for castration, 13, 15, 29
Local anaesthesia for castration, Apps. I and II
    ovariectomy in cattle, 185
    penile relaxation, App. III
London method of casting horses, App. I
Longitudinal canals of bovine penis, 89
Lymphatic drainage of inguinal region, 57, 73

M

Mare, caesarean anaesthesia, 153
    , for uterine torsion, 161
    , indications for, 145, 146
    , post-operative care, 163 - 167
    , technique for, 155, 156, 157, 158, 160
    , Caslick's operation in, 188
    , ovariectomy in, 183 - 185
    , perineal damage in, 171
    , Pouret's operation in, 188
    , tumours on clitoris of, 81
    , urine pooling in, 187, 189
Maternal dystocia as an indication for caesarean, 146
Metritis after caesarean operation, 165
    due to pneumovagina, 176, 186
Monorchid, 37

Monsters, as an indication for caesarean, 146, 160
Mummification, as an indication for caesarean, 146
Muscles, bulbospongiosus, 71, 72, 87
    cremaster, 6, 7, 9, 13, 45, 46
    ischio-cavernosus, 71, 72, 87, 91, 114
    body wall, in caesarean, 155, 159
    cryptorchidectomy, 47
    of inguinal canal, 54 - 57
    preputial, 92 - 93, 111
    retractor penis, 71, 88, 89, 97, 110-111
    urethral, 71

N

Neoplasia, see Tumours
New born, care of, 163
Nymphomania in mares, 183

O

Oedema after caesarean, 166
    castration, 17, 26, 30, 34
    cryptorchidectomy, 49
    of penis, 70, 77, 78, 79
Omentum, prolapse of, after castration, 17, 63
Orchitis, 33 - 34
Ovariectomy, 183 - 186
Ovary, adhesions of, 158, 167
    , cystic, 183
    , tumours of, 183
Oxytocin after caesarean, 160, 165
    as an alternative to caesarean, 149
    in hydrallantois, 163

P

Paralysis of horse's penis, 77
Paraphimosis, 76, 77, 83, 100
Paravertebral anaesthesia, App. II
Parturition difficulties - see Dystocia
    , induction of, 149
    , injuries at, 171 - 181
    , prediction of, 146
Pelvic ligaments, 127
Pelvic symphysiotomy versus caesarean, 148
Pelvic urethra, 88
Penile adhesions, 95, 104, 105, 107
    frenulum, 98 - 99
    integument, 92 - 93
    part of prepuce, 92
    rupture - see Penis, haematoma of
Penis, amputation of, 84, 119
    , anatomy of, 71 - 73, 87 - 93
    , deviation of, 100 - 102, 119 - 121
    , examination of, 74, 94 - 96
    , haematoma of, 77, 103, 106, 107
    , infantile, 97
    , insufficient protrusion of, 94 - 96
    , paralysis of, 77
    , removal of, at castration, 20
    , retractor, muscle of, 71, 88, 89, 97, 110 - 111
    - see also Balanoposthitis, Prepuce
    , traumatic lesions of, 77, 103 - 108
    , tumours of, 81, 109
Perineum, damage at parturition to, 171, 190
Peritonitis after caesarean, 166
    castration, 19, 30
    cryptorchidectomy, 49
    ovariectomy, 185
    trauma to reproductive tract, 180, 181
Pessaries, antibiotic, after caesarean, 153, 158, 165
Phimosis, 100
Piglet, castration of, 29, App. II
    , inguinal hernia in, 60, 65
Pneumovagina in mares, 186
Polyorchid, 37
Post-partum haemorrhage, 172, 180
    , injuries, 171 - 182

Pouret's operation for pneumovagina, 188
Pre-penile part of the prepuce, 92
Prepuce, adhesions within, 74, 83, 94 - 96, 107, 111 - 112
    , anatomy of, 72, 92 - 93
    , eversion, 111
    , infection of, 107, 111
    , lacerations of, 77, 107, 111,
    , prolapse of, 77, 103, 111 - 113
    , stenosis, 111 - 112
Preputial fascia, rupture of, 81
Prolapse of bowel following castration, 17, 63
    omentum following, 17, 63
    penis, 77
    prepuce, 77, 103, 111 - 113
    vagina, 127 - 144
Prolonged pregnancy, 146
Pudendal nerve block, 96, 97, 109, App. III

R

Rainbow deviation, 102
Ram, castration of, 23, 24, 29, App. II
    , cryptochidism in, 37
    , inguinal hernia in, 61, 62
    , penis of, 87, 90, 107
    , vasectomy in, 119, 120
Recto-vestibular fistula, 173, 174
    tears, 171, 173 - 177
Reefing operation for equine penis, 83
Retraction of equine penis, failure of, 77
    , normal, 71
    , surgery for, 83
    of bovine penis, normal, 91
Rig, definition of, 37
    , false, 43
    , origin of term, 37
    - see also cryptorchid

S

Schistosomus reflexa, 146, 160
Scirrhous cord, 19, 30
Scrotal hernia - see Inguinal hernia
Scrotal swelling in gelding, 17 - 20, 33, 61
    stallion, 32, 33, 62, 64
Scrotum, abscess in, 18, 19, 30, 33
    , anatomy of, 5
    , cysts in, 33
    , haematoma in, 17, 33
Sheath - see Prepuce
Sigmoid flexure, 87, 91, 97, 103
Slaughter versus caesarean, 150
Social behaviour of teaser animals, 124
Sow, caesarean in, 152, 155, 157, 159, 164
    , management of, at castration of piglets, 29
    , prolapse of the vagina in, 133 - 135
Spiral deviation of bovine penis, 100
Stallion, castration of, 10 - 15
    , inguinal hernia in, 62
    , inguinal rupture in, 64
    , penis of, 71
    , vasectomy of, 119
Standing castration, App. I
Strangulated hernia or rupture, 53, 62, 64, 67 - 69
Support for prolapsed penis, 80
    prepuce, 113
Suture abscess after castration, 19
Symphysiotomy versus caesarean, 148

T

Teaser animals, choice of, 120
    , management of, 124
    , preparation of, 120
Teratoma of testis, 32, 40, 48
Testis, absence of, 37
    , anatomy of, 5 - 9

, coverings of, 5 - 6
, cystic, 32, 48
, descent of, 1 - 5
, ectopic, 37
, inflammation of, 33
, removal of - see Castration
, retention of - see Cryptorchid
, tumours of, 32
Thiopentone not to be used for caesarean, 153
Torsion of spermatic cord, 34
Training, castration of horses in, 12
Trauma as a cause of hernia or rupture, 61
       to bovine penis, 103 - 108
          equine penis, 77
          testis and scrotum, 33
          vagina, vestibule and vulva, 171 - 181
Tumours of bovine penis, 109
        equine penis, 81
        equine ovary, 183
        testis, 32
        vulva and clitoris, 81

U

Umbilical cord at caesarean, 158
Undescended testis, diagnosis of, 42, 43
                , palpation of, 42, 45
                , treatment of, 45 - 48
                - see also Cryptorchidism
Uterus, haemorrhage from, 158, 178, 180
      , incising at caesarean, 157
      , inflammation of - see Metritis
      , tears in, 177
Urethra of bull, 35, 87, 108
      horse, 71, 72, 82, 84
Urethral blockage after castration, 24, 26
      extension for urine pooling, 189
      process, 72
Urethritis in bull, 108
           stallion, 82

V

Vaginal process, 1, 6, 8, 12, 13, 28, 33, 45, 53, 57
      ring, 1, 11, 58, 61, 63, 65
      tunic, 1, 16, 45, 61, 62, 65, 67
Vagina/Vestibule, abscesses in, 172
            , anatomy of, 127 - 129, 171
            , annular constriction between, 133
            , cysts in, 192
            , haematoma of, 172
            , prolapse of, 127 - 144
            , tears in, 172
            , varicosity, 193
Vaginitis and cysts in vestibular gland, 193
         in pneumovagina, 186
            recto-vestibular injuries, 176
Varicocele, 32
Varicosity, Vaginal, 193
Vas deferens, anatomy of, 7, 9
Vasectomy, interval to use after, 124
          , management after, 124
          , species suitable for, 119
          , surgical technique for, 120
Vesicular gland of bull, 88
Vestibular glands, 192
Vestibule, distinguish from vulva, 127, 189
      - see also Vagina/Vestibule, Vulva
Vulva, haematoma in, 172
      , tears in, 172, 190